Managing Water and Agroecosystems for Food Security

Comprehensive Assessment of Water Management in Agriculture Series

Titles Available

Managing Water and Agroecosystems for Food Security

Edited by

Eline Boelee

Water Health, Hollandsche Rading, the Netherlands
International Water Management Institute, Colombo, Sri Lanka

www.cabi.org

CABI is a trading name of CAB International

CABI Head Office
Nosworthy Way
Wallingford
Oxfordshire OX10 8DE
UK

Tel: +44 (0)1491 832111
Fax: +44 (0)1491 833508
E-mail: info@cabi.org
Website: www.cabi.org

CABI
38 Chauncey Street
Suite 1002
Boston, MA 02111
USA

T: +1 800 552 3083 (toll free)
T: +1 (0)617 395 4051
E-mail: cabi-nao@cabi.org

A catalogue record for this book is available from the British Library, London, UK.

Library of Congress Cataloging-in-Publication Data
Managing water and agroecosystems for food security / edited by Eline Boelee.
     p. cm. -- (Comprehensive assessment of water management in agriculture series ; 10)
  ISBN 978-1-78064-088-4 (alk. paper)
1. Agricultural ecology. 2. Water in agriculture. 3. Food security. I. Boelee, Eline. II. Series: Comprehensive assessment of water management in agriculture series ; 10.

 S589.7.M36 2013
 577.5'5--dc23

2013002767

ISBN-13: 978 1 78064 088 4

Commissioning editor: Victoria Bonham
Editorial assistant: Alexandra Lainsbury
Production editor: Lauren Povey

Typeset by Columns Design XML Ltd, Reading, UK.
Printed and bound in the UK by CPI Group (UK) Ltd, Croydon, CR0 4YY.

Contents

Contributors

(Note that e-mail addresses are included for lead authors only.)

T. Amede, *International Crops Research Institute for the Semi-arid Tropics (ICRISAT), PO Box 1906, Maputo, Mozambique. (formerly CGIAR Challenge Program on Water and Food – CPWF and International Water Management Institute/International Livestock Research Institute – IWMI/ILRI).*

S. Barchiesi, *International Union for Conservation of Nature (IUCN), Global Water Programme, Rue Mauverney 28, 1196 Gland, Switzerland.*

J. Barron, *Stockholm Environment Institute, University of York, Heslington, York, YO10 5DD, UK and Stockholm Resilience Centre, Stockholm University, Stockholm, Sweden. E-mail: jennie.barron@sei-international.org*

M. Beveridge, *WorldFish (Lusaka Office), PO Box 51289, Ridgeway, Lusaka, Zambia.*

P. Bindraban, *World Soil Information (ISRIC) and Plant Research International, Wageningen University and Research Centre (Wageningen UR), PO Box 353, 6700 AJ Wageningen, the Netherlands.*

E. Boelee, *Water Health, Tolakkerweg 21, 3739 JG Hollandsche Rading, the Netherlands (formerly IWMI). E-mail: e.boelee@waterhealth.nl*

S.W. Bunting, *Essex Sustainability Institute, School of Biological Sciences, University of Essex, Wivenhoe Park, Colchester, Essex, CO4 3SQ, UK. E-mail: swbunt@essex.ac.uk*

F. Clement, *IWMI, GPO Box 8975, EPC 416, Kathmandu, Nepal.*

D. Coates, *Secretariat of the Convention on Biological Diversity (CBD), 413, Saint Jacques Street, suite 800, Montreal QC H2Y 1N9, Canada. E-mail: david.coates@cbd.int*

S. Cook, *CGIAR Research Program on Water, Land and Ecosystems, PO Box 2075, Colombo, Sri Lanka.*

K. Descheemaeker, *Plant Production Systems, Wageningen University, Droevendaalsesteeg 1, 6708 PB Wageningen, the Netherlands (formerly Commonwealth Scientific and Industrial Research Organisation – CSIRO and IWMI/ILRI). E-mail: katrien.descheemaeker@ wur.nl*

N. Eriyagama, *IWMI, PO Box 2075, Colombo, Sri Lanka.*

A. Evans, *Edge Grove School, Aldenham Village, Watford, WD25 8NL, UK (formerly IWMI).*

M. Finlayson, *Institute for Land, Water and Society (ILWS), Charles Sturt University, PO Box 789, Albury, NSW 2640, Australia. E-mail: mfinlayson@csu.edu.au*

R. Fleiner, *International Centre for Integrated Mountain Development (ICIMOD), GPO Box 3226, Khumaltar, Kathmandu, Nepal. E-mail: rfleiner@icimod.org*

D. Grace, *ILRI, PO Box 30709, Nairobi 00100, Kenya.*

M. Herrero, *CSIRO, 306 Carmody Road, St Lucia, Queensland 4067, Australia; formerly ILRI.*

D.I. Jarvis, *Bioversity International, Via dei Tre Denari 472/a, 00057 Maccarese (Fiumicino) Rome, Italy. E-mail: d.jarvis@cgiar.org*

E. Khaka, *United Nations Environment Programme (UNEP), PO Box 30552, 00100, Nairobi, Kenya. E-mail: elizabeth.khaka@unep.org*

L. Korsgaard, *UNEP–DHI Centre for Water and Environment, Agern Allé 5, DK 2970 Hørsholm, Denmark. E-mail: lok@dhigroup.com*

G.J. Lloyd, *UNEP–DHI Centre for Water and Environment, Agern Allé 5, DK 2970 Hørsholm, Denmark. E-mail: gjl@dhigroup.com*

J.C. Milder, *EcoAgriculture Partners, 1100 17th Street NW, Suite 600, Washington, DC 20036, USA.*

D. Molden, *ICIMOD, GPO Box 3226, Khumaltar, Kathmandu, Nepal (formerly IWMI).*

C. Muthuri, *World Agroforestry Centre (ICRAF), United Nations Avenue, Gigiri, PO Box 30677, Nairobi 00100, Kenya.*

S. Nguyen-Khoa, *World Water Council (WWC), Espace Gaymard, 2–4 Place d'Arvieux, 13002 Marseille, France (formerly CPWF).*

P.L. Pert, *CSIRO, Ecosystem Sciences, PO Box 12139, Cairns, Queensland 4870, Australia. E-mail: petina.pert@csiro.au*

L. Sanford, *IWMI, PO Box 2075, Colombo, Sri Lanka.*

S.J. Scherr, *EcoAgriculture Partners, 1100 17th Street NW, Suite 600, Washington, DC 20036, USA.*

E.M. Solowey, *The Arava Institute for Environmental Studies (AIES), Kibbutz Ketura, D.N. Hevel Eilot 88840, Israel. E-mail: elaine.solowey@arava.org*

R.E. Tharme, *The Nature Conservancy (TNC), 48 Middle Row, Cressbrook, Buxton, Derbyshire, SK17 8SX, UK. E-mail: rtharme@tnc.org*

L. Thiombiano, *Central Africa Bureau, Food and Agriculture Organization of the United Nations (FAO), BP 2643, Libreville, Gabon.*

M. van Brakel, *CGIAR Research Program on Water, Land and Ecosystems, PO Box 2075, Colombo, Sri Lanka (formerly CPWF).*

Series Foreword:
The Comprehensive Assessment (CA) of Water Management in Agriculture

There is broad consensus on the need to improve water management and to invest in water for food, as these are critical to meeting the Millennium Development Goals (MDGs). The role of water in food and livelihood security is a major issue of concern in the context of persistent poverty and continued environmental degradation. Although there is considerable knowledge on the issue of water management, an overarching picture on the water–food–livelihoods–environment nexus is missing, leaving uncertainties about management and investment decisions that will meet both food and environmental security objectives.

The Comprehensive Assessment (CA) of Water Management in Agriculture is an innovative multi-institute process aimed at identifying existing knowledge and stimulating thought on ways to manage water resources to continue meeting the needs of both humans and ecosystems. The CA critically evaluates the benefits, costs and impacts of the past 50 years of water development and the challenges to water management currently facing communities. It assesses innovative solutions and explores the consequences of potential investment and management decisions. The CA is designed as a learning process, engaging networks of stakeholders to produce knowledge synthesis and methodologies. The main output of the CA is an assessment report that aims to guide investment and management decisions in the near future, considering their impact over the next 50 years in order to enhance food and environmental security to support the achievement of the MDGs. This report was published in 2007 under the title *Water for Food, Water for Life: A Comprehensive Assessment of Water Management in Agriculture*, but CA research and knowledge-sharing activities have continued to expand the assessment.

The primary assessment research findings are presented in a series of books that forms the scientific basis of the CA of Water Management in Agriculture. The books cover a range of vital topics in the areas of water, agriculture, food security and ecosystems – the entire spectrum of developing and managing water in agriculture, from fully irrigated to fully rainfed lands. They are about people and society, why they decide to adopt certain practices and not others and, in particular, how water management can help poor people. They are about ecosystems – how agriculture affects ecosystems, the goods and services that ecosystems provide for food security, and how water can be managed to meet both food and environmental security objectives. This is the tenth book in the series.

Managing water effectively to meet food and environmental objectives will require the concerted action of individuals from across several professions and disciplines – farmers, fishers, pastoralists, water managers, economists, hydrologists, irrigation specialists, agronomists and social scientists. The material presented in this book represents an effort to synthesize recent research building on the CA and proposes an ecological approach to food security, where agroecosystems, water resources and other landscape elements are managed together at landscape level. The complete set of books should be invaluable for resource managers, researchers and field implementers. These books will provide source material from which policy statements, practical manuals and education and training materials can be prepared.

The CA has been carried out by a coalition of partners that includes 11 Future Harvest agricultural research centres supported by the CGIAR (originally so named as the acronym for the Consultative Group on International Agricultural Research), the Food and Agriculture Organization of the United Nations (FAO) and partners from some 80 research and development institutes globally. Co-sponsors of the assessment – institutes that are interested in the results and help frame the assessment – are the Ramsar Convention, the Convention on Biological Diversity (CBD), the FAO and the CGIAR.

For the production of this book, financial support from the Swiss Agency for Development and Cooperation (SDC) and the National Institute for Rural Engineering in Japan is appreciated. Development of content has been facilitated by financial and logistical support from the United Nations Environment Programme (UNEP).

David Molden
Series Editor up to Volume 9
Formerly of International Water
Management Institute
Sri Lanka

Series Foreword:
Water, Land and Ecosystems

As we move into the era of the Anthropocene, in which human actions have become the main driver of global environmental change, there is clear evidence to support the notion that the earth's systems have been pushed outside the stable state, with consequences that could have irreversible and, in some cases, abrupt environmental change, so leading to a state less conducive to human development. Our agricultural production systems, which, so far, have successfully provided food, feed and fibre to an ever-increasing global population, are based on the insatiable consumption of fertilizers, a dependence on fossil fuels and massive changes in land use that have contributed to increasing greenhouse gases in our atmosphere, and to loss of biodiversity and mass species extinction never before seen in human history.

It is clear that there is a need to change the way we do business in the agricultural sector if we are to adequately provide food, feed and fibre to a global population that is destined to peak at 9 billion in 2050. The contributors to this book, Volume 10 of the *Comprehensive Assessment of Water Management in Agriculture Series*, are enthusiastically optimistic that we can achieve this through a paradigm shift that places agriculture within an ecosystem context that is more efficient in its use of natural resources and promotes the provisioning of ecosystem services. This is the first of what is hoped to be many outputs from the newly formulated Research Program on Water, Land and Ecosystems (WLE) of the CGIAR.

The WLE Research Program builds on the findings from the completed Comprehensive Assessment (CA) of Water Management in Agriculture process by seeking a paradigm shift that views the sustainable management of multifunctional landscapes as the most cost-effective strategy to boost agricultural production, improve livelihoods, increase food security and alleviate poverty. The programme's goal is to achieve sustainable improvements in agricultural productivity required to produce enough food for all and generate sufficient income to lift millions of smallholder households from poverty, while ensuring their food and nutritional security.

The book provides a synthesis of existing knowledge on ways to manage water and agroecosystems that enhance nature's services beyond food production, while identifying areas for further research. It pays specific attention to the impacts of agricultural water management on ecosystems, and the importance of ecosystems in supporting water for agriculture. In so doing, it sets the stage for addressing the main and overarching research questions of the WLE Research Program, namely, how to: (i) reverse land degradation; (ii) address water scarcity; and (iii) achieve both agricultural intensification and the enhancement of a broad range of ecosystem services.

Andrew Noble
CGIAR Research Program on Water, Land and Ecosystems
Sri Lanka

Acknowledgements

This book is largely a synthesis of existing knowledge and gaps therein, from international references as well as from experts. However, the authors have based their insights on various research projects in which they have participated over recent years. We want to express our gratitude to the various site teams and the farmers involved in this research, as well as to many national programmes throughout the globe.

The lead authors (which are indicated by the presence of their e-mail addresses in the Contributors' list and at the heads of the relevant chapters) have been able to develop the content of their chapters with in-kind support of their respective organizations, which we would like to acknowledge here. Hence, for the various authors, we appreciate support from: the International Water Management Institute (IWMI), the CGIAR Research Program on Water, Land and Ecosystems (WLE), as well as Water Health, for Eline Boelee; the Stockholm Environment Institute (SEI), for Jennie Barron; The Nature Conservancy (TNC), for Rebecca Tharme; the Secretariat of the Convention on Biological Diversity (CBD), for David Coates; the Commonwealth Scientific and Industrial Research Organisation (CSIRO), for Petina Pert; the International Centre for Integrated Mountain Development (ICIMOD), for Renate Fleiner; the Arava Institute for Environmental Studies (AIES), for Elaine Solowey; the

European Commission Seventh Framework Programme (FP7) HighARCS (Highland Aquatic Resources Conservation and Sustainable Development) project and the University of Essex, for Stuart Bunting; the Institute for Land, Water and Society, Charles Sturt University, for Max Finlayson; Wageningen University and CSIRO, for Katrien Descheemaeker; the International Fund for Agricultural Development (IFAD), the Global Environmental Facility (GEF), the Swiss Agency for Development and Cooperation (SDC) and Bioversity International, for Devra Jarvis; the United Nations Environment Programme (UNEP), for Elizabeth Khaka; and the UNEP-DHI Centre for Water and Environment, for both Gareth James Lloyd and Louise Korsgaard. The same holds true for all co-authors, who contributed their time and expertise supported by their organizations. Some lead authors and co-authors changed employers during the time this book was prepared, and we explicitly want to recognize all respective organizations.

Most lead authors and several of the co-authors have contributed to other chapters than those where they are listed. As the editor, I want to express my appreciation to all of you who provided inputs and made this book a truly joint and interdisciplinary effort.

In addition to the lead authors and co-authors, many others contributed background settings and feedback at earlier stages of developing the chapters for this book. We sincerely appreciate inputs and support from: Marc Andreini (Daugherty Water for Food Institute, University of Nebraska); Sithara Attapattu; Maija Bertule (UNEP–DHI); Luna Bharati (IWMI); Marta Ceroni (Gund Institute for Ecological Economics, University of Vermont); Thomas Chiramba (UNEP); Karen Conniff; Helen Cousins; Jan de Leeuw (International Livestock Research Institute – ILRI and World Agroforestry Centre – ICRAF); Kristina Donnelly (AIES); Pay Drechsel (IWMI); Mark Giordano (IWMI); Line Gordon (Stockholm Resilience Centre); Clive Lipchin (AIES); Abby Lutman (AIES); Matthew McCartney (IWMI); Bertha Nherera (Pelum); An Notenbaert (ILRI); Tim Pagella (Bangor University); Don Peden (ILRI); Asad Qureshi (IWMI); Fergus Sinclair (ICRAF); Katherine Snyder (IWMI); David Stentiford; Martin van Brakel (CGIAR Challenge Program on Water and Food – CPWF/WLE), Jeanette van de Steeg (ILRI), Gerardo E. van Halsema (Irrigation and Water Engineering, Wageningen University and Research Centre – Wageningen UR); and Kees van 't Klooster (Alterra, Wageningen UR).

The chapters in this book have been reviewed by Peter Hazell (International Food Policy Research Institute – IFPRI); Terry Hills (Conservation International, USA); Robyn Johnston (IWMI); Netij Ben Mechlia (Institution de la Recherche et de l'Enseignement Supérieur Agricoles – IRESA, Tunisia); Siwa Msangi (IFPRI); Francis Murray (Stirling University, UK); Ephraim Nkonya (IFPRI); Markos Tibbo (Food and Agriculture Organization of the United Nations – FAO); and Nick van de Giesen (Delft University of Technology – TU Delft, Netherlands). Several reviewers commented on more than one chapter. With these comments, the lead authors have been able to substantially improve on the contents of the chapters. Subsequently, the book as a whole has been reviewed by Ania Grobicki (Global Water Partnership – GWP); David Lehrer (AIES); Laurence Smith (School of Oriental and African Studies – SOAS, University of London, UK) and Dennis Wichelns (IWMI). These comments challenged us all to further develop, condense and finalize the book.

C.T. Hoanh (IWMI) merits special recognition as he came up with the idea of preparing this book in the CA series, and continued to give his support throughout the process. Without him, this volume would not exist. So thank you, Hoanh, and good luck with your new series!

We hope that the book will contribute to the further development of the paradigm shift towards an ecosystem approach to food production, leading to the wider application of improved management of both ecosystems and water, as well as more in-depth action research on the potential of agroecological landscapes to help provide sustainable food security.

Eline Boelee
Editor

1 Introduction

Eline Boelee,[1]* David Coates,[2] Elizabeth Khaka,[3] Petina L. Pert,[4] Lamourdia Thiombiano,[5] Sara J. Scherr,[6] Simon Cook[7] and Luke Sanford[8]

[1]*Water Health, Hollandsche Rading, the Netherlands;* [2]*Secretariat of the Convention on Biological Diversity (CBD), Montreal, Canada;* [3]*United Nations Environment Programme (UNEP), Nairobi, Kenya;* [4]*Commonwealth Scientific and Industrial Research Organisation (CSIRO), Cairns, Queensland, Australia;* [5]*Central Africa Bureau, Food and Agriculture Organization of the United Nations (FAO), Libreville, Gabon;* [6]*EcoAgriculture Partners, Washington, DC, USA;* [7]*CGIAR Research Program on Water, Land and Ecosystems, Colombo, Sri Lanka;* [8]*International Water Management Institute (IWMI), Colombo, Sri Lanka*

Abstract

This chapter sets the stage for our book on *Managing Water and Agroecosystems for Food Security*. It provides an introduction to the extent of food insecurity in the world and how this is further jeopardized by unsustainable food production. Water is a main constraint to sustainability because water use in agriculture has huge impacts on downstream ecosystems. Furthermore, degraded ecosystems are less capable of sustaining water flows. In this book the authors take an ecosystem approach to freshwater management for sustainable agroecosystems and food security, with an emphasis on technical options. They show how water and ecosystems can be managed in such a way that they are mutually supportive and contribute to sustainable food security and wealth.

Background

The global food shortages and soaring food prices of the 2000s led to increased attention to food security worldwide. Rising food prices are continuously aggravated by population growth and climatic factors. Globally, about 870 million people, mostly from developing countries, are undernourished (FAO *et al.*, 2012). Most of these people live in countries that are not self-sufficient in food production, in particular in South Asia and sub-Saharan Africa, where agricultural productivity is often low. This is due to factors such as limited soil nutrient availability, the occurrence of pests and diseases, and spells of minimal or no precipitation or irrigation during critical growing periods. Poor agricultural practices have aggravated land degradation so that it is now seriously limiting food production (Bossio and Geheb, 2008).

*E-mail: e.boelee@waterhealth.nl

Fisheries and aquaculture, which are major sources of protein in many developing countries, provided more than 2.9 billion people with at least 15% of their average per capita animal protein intake in 2006 (FAO, 2009b), but these too are threatened by ecosystem degradation caused by overfishing, habitat destruction, pollution, invasive species and the disruption of river flow by dams. These pressures have caused a severe decline in fish species diversity and production, particularly in inland fisheries, thus threatening an important food and nutrition source for low-income rural men, women and children (UNEP, 2010). Beef, poultry, pork and other meat products provide one third of humanity's protein intake but also consume almost a third (31%) of the water used in agriculture globally (Herrero et al., 2009).

Agriculture and ecosystem services are interrelated in various ways. Agroecosystems generate beneficial ecosystem services such as the production of food, feed and fibre, but they also generate biodiversity, carbon storage, water services, soil retention and aesthetic benefits (Wood et al., 2000; UNEP, 2007). In return, agroecosystems receive beneficial ecosystem services from other ecosystems, such as pollination and a supply of fresh water. However, ecosystem services from non-agricultural systems may be affected by agricultural practices and, in turn, dysfunctional ecosystem services have further impacts on agroecosystems and their production systems, thereby threatening food security (Hassan et al., 2005; Millennium Ecosystem Assessment, 2005a, 2005b; Nellemann et al., 2009).

These various environmental pressures on, and negative trends in, food production are further threatened by climate change (see Chapter 2 for more detailed discussion). Increases in the magnitude and frequency of drought and floods are expected to lead to higher spatial and temporal variability in production and lower overall food production, especially in sub-Saharan Africa (Parry et al., 2007).

Feeding a world population of over 9 billion people in 2050 will require the raising of overall food production by some 70% over the period from 2005–2007 to 2050 (nearly 100% in low-income countries) (FAO, 2009a), in addition to the putting in place of global and national mechanisms to ensure equitable access. Obviously, food security is not only a matter of food production but also an issue of equity and secure access to the means of production and to food products (FAO, 2010). Thus, food security is the product of many variables, which include: physical factors such as climate, soil type and water availability; the management of these factors and of other natural resources (water, land, aquatic resources, trees and livestock), at the level of fields, landscapes and river basins; and losses and waste along the value chain (see Chapter 2). Food security requires supporting policies to ensure more equitable access to food, while agroecosystems have to be managed in a more sustainable way so as to increase long-term food security and livelihood benefits while minimizing or reversing environmental deterioration.

The understanding of linkages between ecosystems, water and food production is important to the health of all three, and managing for the sustainability of these connections is becoming increasingly necessary to help in improving global food security (Molden, 2007). Changes in the global water cycle, caused largely by human pressures, are seriously affecting ecosystem health and human well-being (Millennium Ecosystem Assessment, 2005c; WWAP, 2012; see Chapter 5). For example, in key parts of the tropics, agriculture has continued to expand into forest and woodland areas (Gibbs et al., 2010), where it has caused reduced tree cover and soil compaction, which have led to reduced infiltration and higher runoff of rainwater, often causing severe erosion, salinization or other degradation processes (Ong and Swallow, 2003; Falkenmark et al., 2007). Ecosystem degradation therefore threatens the regulation of ecosystem services such as water quality and water flow. Likewise, water is a key driver of several ecosystem functions, including biomass and crop yields, as well as of various supporting and regulatory ecosystem services (Keys et al., 2012).

To address the significant sustainability issues in agriculture, particularly that of water use, the agricultural sector needs the development and implementation of a functioning ecosystems approach to water management and food security. This in turn helps to increase productivity, i.e. it produces more, and better, food without further increase in the use of land, water and other valuable inputs, particularly in sub-Saharan Africa and other vulnerable regions. Global assessments suggest that despite the planetary limits to resource availability, it is feasible to achieve sustainable agricultural production while simultaneously meeting other human needs, although this requires significant changes in policy and approach (Foley *et al.*, 2011). Increased water productivity is crucial to achieving sustainable food security (Fisher and Cook, 2010).

Potential of Ecosystem Approach

The challenges to food security can be addressed by managing agriculture as eco-systems that require certain water flows and provide essential ecosystem services, supported by appropriate policy and institutions. In practical terms this would mean improving agricultural management across scales (from field to landscape or basin level), linking to downstream aquatic ecosystems, and creating and managing multifunctional agroecosystems (Gordon *et al.*, 2010). In this book, we define agroecosystems as a set of human practices, aimed at food production – and embedded in and part of its own ecosystem – that has certain ecosystem needs, functions and services, and that interacts with other natural and human-made ecosystems (see Chapter 3). Agroecosystem management is then the management of natural resources and of other inputs for the sustainable production of food and of other provisioning, cultural, regulatory and supporting ecosystem services (see Chapter 4).

One of the shaping characteristics of an agroecosystem is its climate, which helps to determine the length of the available growth period (LEAD, 1999). In tropical areas four zones are distinguished: arid, semi-arid, sub-humid and humid. In temperate regions and highlands the mean monthly temperature is the main determinant of the climate. The particularly fragile arid zone and its challenges are discussed in more detail in Chapter 6. Wetlands are found across all zones and provide many high-value ecosystem services, which is why they are increasingly exploited for, and threatened by, food production (see Chapter 7).

While a paradigm shift towards an ecosystems-based approach to water and food security has begun (UNEP, 2011; Frison, 2012; Keys *et al.*, 2012; Landscapes for People, Food and Nature Initiative, 2012; WLE, 2012), it is vitally important to continue the application of this to what we already know and to encourage innovations in the approach. Hence, in this volume the authors show how ecosystems and water can be managed in such a way that they mutually support food production, thereby contributing to sustainable food security. The book illustrates the three-way interdependence between ecosystems, water management and food security (Fig. 1.1). By looking at the world as a range of interlinked ecosystems (from naturally pristine to the highly intensive agriculture of crops, livestock, fish and trees) and recognizing the variety of ecosystem services, the improved management of water and ecosystems together has the potential to bring long-term food security.

The book is structured to systematically show the relationships between ecosystems, water and food security, and to elaborate an ecosystem approach to sustainable agriculture. It contains chapters on the drivers of food security (Chapter 2) and provides solid analyses on ecosystems, agroecosystems, ecosystem services and their valuation (Chapters 3 and 4). Next, there is an analysis of the role of water in agriculture as well as analyses of water use and scarcity (Chapter 5). This is followed by discussions of the specific challenges in drylands (Chapter 6) and wetlands (Chapter 7); each of these chapters provides more insight into the reasons why an integrated ecosystem approach is required and what this should entail, giving practical recommendations for those vulnerable ecosystems. A discussion of the contributions that can be made by increased water productivity to a better joint management of agroecosystems and water follows in the next chapter (Chapter 8). Subsequently, Chapter 9 presents various approaches to the enhancement of ecosystem services in agri-culture, with many concrete examples, while

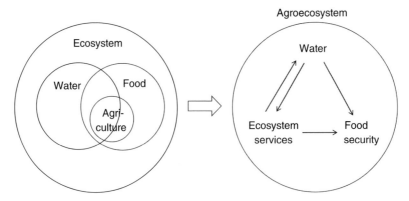

Fig. 1.1. Water and food as dimensions of ecosystems (left), with agriculture as a subset of food (production), and the role of water for food security and other ecosystem services in an agroecosystem (right).

Chapter 10 provides more detail of the ecosystem approach to water management. Finally, the last chapter (Chapter 11) ends the book with a synthesis that embeds the key recommendations into a landscape approach, links this to ongoing initiatives and identifies knowledge gaps for further research.

Conclusions

With a growing global population expected to reach around 9 billion in 2050, and the increasing impacts of climate change, the sustainable use of water and ecosystems for food security is a great challenge. It has become increasingly important to gain a better understanding of the functioning of terrestrial and aquatic ecosystems, and their interrelations with the availability and quality of water. This calls for a shift in the management of ecosystems and the water within them for food security. Ecosystems need to be safeguarded and the resources within used wisely, as they are the backbone of all environmental services needed in achieving food security and are often of direct importance to low-income countries and marginal groups.

References

Bossio, D. and Geheb, K. (eds) (2008) *Conserving Land, Protecting Water.* Comprehensive Assessment of Water Management in Agriculture Series 6. CAB International, Wallingford, UK, in association with CGIAR Challenge Program on Water and Food, Colombo and International Water Management Institute (IWMI), Colombo.

Falkenmark, M., Finlayson, C.M., Gordon, L.J. *et al.* (2007) Agriculture, water, and ecosystems: avoiding the costs of going too far. In: Molden, D. (ed.) *Water for Food, Water for Life: Comprehensive Assessment of Water Management in Agriculture.* Earthscan, London, in association with International Water Management Institute (IWMI), Colombo, pp. 233–277.

FAO (2009a) *Global Agriculture Towards 2050. How to Feed the World in 2050: High Level Expert Forum, Rome 12–13 October 2009.* Issues Paper HLEF2050, Food and Agriculture Organization of the United Nations, Rome. Available at: http://www.fao.org/fileadmin/templates/wsfs/docs/Issues_papers/HLEF2050_Global_Agriculture.pdf (accessed February 2013).

FAO (2009b) *The State of World Fisheries and Aquaculture 2008.* Food and Agriculture Organization of the United Nations, Rome.

FAO (2010) *The State of Food Insecurity in the World. Addressing Food Insecurity in Protracted Crises.* Food and Agriculture Organization of the United Nations, Rome.

FAO, WFP and IFAD (2012) *The State of Food Insecurity in the World 2012*. Jointly published by Food and Agriculture Organization of the United Nations, World Food Programme and International Fund for Agricultural Development, Rome. Available at: http://www.fao.org/docrep/016/i3027e/i3027e.pdf (accessed February 2013).

Fisher, M. and Cook, S. (2010) Introduction: Special issue: Water, food and poverty in river basins, Part 1. *Water International* 35, 465–471. doi:10.1080/02508060.2010.520223

Foley, J.A. *et al.* (2011) Solutions for a cultivated planet. *Nature* 478, 337–342. doi:10.1038/nature10452

Frison, E. (2012) Bringing conservation and agriculture together. Bioversity International, Rome. Available at: www.bioversityinternational.org/index.php?id=6650 (accessed September 2012).

Gibbs, H.K., Ruesch, A.S., Achard, F., Clayton, M.K., Holmgren, P., Ramankutty, N. and Foley, J.A. (2010) Tropical forests were the primary sources of new agricultural land in the 1980s and 1990s. *Proceedings of the National Academy of Sciences of the United States of America* 107, 16732–16737. doi:10.1073/pnas.0910275107

Gordon, L.J., Finlayson, C.M. and Falkenmark, M. (2010) Managing water in agriculture for food production and other ecosystem services. *Agricultural Water Management* 94, 512–519. doi:10.1016/j.agwat.2009.03.017

Hassan, R., Scholes, R. and Ash, N. (eds) (2005) *Ecosystems and Human Well-being: Current State and Trends, Volume 1. Findings of the Condition and Trends Working Group of the Millennium Ecosystem Assessment.* World Resources Institute and Island Press, Washington, DC.

Herrero, M., Thornton, P.K., Gerber, P. and Reid, R.S. (2009) Livestock, livelihoods and the environment: understanding the trade-offs. *Current Opinion in Environmental Sustainability* 1, 111–120. doi:10.1016/j.cosust.2009.10.003

Keys, P., Barron, J. and Lannerstad, M. (2012) *Releasing the Pressure: Water Resource Efficiencies and Gains for Ecosystem Services*. United Nations Environment Programme (UNEP), Nairobi and Stockholm Environment Institute (SEI), Stockholm.

Landscapes for People, Food and Nature Initiative (2012) *Landscapes for People, Food and Nature: The Vision, the Evidence, and Next Steps*. EcoAgriculture Partners on behalf of Landscapes for People, Food and Nature Initiative, Washington, DC.

LEAD (1999) Livestock and Environment Toolbox. Livestock, Environment and Development Initiative, Food and Agriculture Organization of the United Nations (FAO), Rome, Italy. Available at: www.fao.org/ag/againfo/programmes/en/lead/toolbox/Index.htm (accessed February 2012).

Millennium Ecosystem Assessment (2005a) *Ecosystems and Human Well-being: Synthesis. A Report of the Millennium Ecosystem Assessment*. World Resources Institute and Island Press, Washington, DC. Available at: www.maweb.org/documents/document.356.aspx.pdf (accessed February 2013).

Millennium Ecosystem Assessment (2005b) *Ecosystems and Human Well-being: Wetlands and Water – Synthesis. A Report of the Millennium Ecosystem Assessment*. World Resources Institute, Washington, DC. Available at: www.maweb.org/documents/document.358.aspx.pdf (accessed February 2013).

Millennium Ecosystem Assessment (2005c) *Living Beyond Our Means. Natural Assets and Human Well-being. Statement from the Board*. Available at: http://www.unep.org/maweb/documents/document.429.aspx.pdf (accessed February 2013).

Molden, D. (ed.) (2007) *Water for Food, Water for Life: Comprehensive Assessment of Water Management in Agriculture*. Earthscan, London, in association with International Water Management Institute (IWMI), Colombo.

Nellemann, C., MacDevette, M., Manders, T., Eickhout, B., Svihus, B., Prins, A.G. and Kaltenborn, B.P. (eds) (2009) *The Environmental Food Crisis – The Environment's Role in Averting Future Food Crises. A UNEP Rapid Response Assessment*. United Nations Environment Programme, GRID-Arendal, Norway. Available at: http://www.grida.no/files/publications/FoodCrisis_lores.pdf (accessed February 2013).

Ong, C.K. and Swallow, B.M. (2003) Water productivity in forestry and agroforestry. In: Kijne, J.W., Barker, R. and Molden, D. (eds) (2003) *Water Productivity in Agriculture: Limits and Opportunities for Improvement*. Comprehensive Assessment of Water Management in Agriculture Series 1. CAB International, Wallingford, UK, in association with International Water Management Institute, Colombo, pp. 217–228.

Parry, M.L., Canziani, O.F., Palutikof, J.P., van der Linden, P.J. and Hanson, C.E. (eds) (2007) *Climate Change 2007: Impacts, Adaptation and Vulnerability. Working Group II Contribution to the Fourth Assessment Report of the Intergovernmental Panel on Climate Change*. Cambridge University Press, Cambridge, UK. Available at: www.ipcc.ch/publications_and_data/ar4/wg2/en/contents.html (accessed May 2011).

UNEP (2007) *Global Environment Outlook. GEO-4, Environment for Development.* United Nations Environment Programme, Nairobi. Available at: www.unep.org/geo/geo4.asp (accessed February 2010).

UNEP (2010) *Blue Harvest: Inland Fisheries as an Ecosystem Service.* WorldFish, Penang, Malaysia and United Nations Environment Programme, Nairobi.

UNEP (2013) Green economy. United Nations Environment Programme, Nairobi. Available at: www.unep. org/greeneconomy (accessed February 2013).

WLE (2012) Agriculture and ecosystems blog. CGIAR Research Program on Water, Land and Ecosystems. Available at http://wle.cgiar.org/blogs/ (accessed September 2012).

Wood, S., Sebastian, K. and Scherr, S.J. (2000) *Pilot Analysis of Global Ecosystems: Agroecosystems. A Joint Study by International Food Policy Research Institute and World Resources Institute.* International Food Policy Research Institute and World Resources Institute, Washington, DC. Available at: www.wri.org/publication/pilot-analysis-global-ecosystems-agroecosystems (accessed January 2012).

WWAP (World Water Assessment Programme) (2012) *The United Nations World Water Development Report 4 (WWDR4): Managing Water Under Uncertainty and Risk (Vol. 1), Knowledge Base (Vol. 2)* and *Facing the Challenges (Vol. 3).* United Nations Educational, Scientific and Cultural Organization (UNESCO), Paris.

2 Drivers and Challenges for Food Security

Jennie Barron,[1]* Rebecca E. Tharme[2]† and Mario Herrero[3]

[1]*Stockholm Environment Institute, University of York, UK and Stockholm Resilience Centre, Stockholm University, Stockholm, Sweden;* [2]*The Nature Conservancy (TNC), Buxton, UK;* [3]*Commonwealth Scientific and Industrial Research Organisation (CSIRO), St Lucia, Queensland, Australia*

Abstract

At the global scale, humanity is increasingly facing rapid changes, and sometimes shocks, that are affecting the security of our food systems and the agroecosystems that are the ultimate sources of food. To plan and prepare for resilient food production and food security in a sustainable and efficient way, we are challenged to better understand the conditions and likely responses of these diverse agroecosystems under various drivers of change and scenarios of future trends. Among the many direct drivers and indirect pressures that exist or are emerging, the discussion in this chapter focuses on the main themes of drivers of demographic changes, globalization of economic and governance systems (including markets), and climate change. The current state of health of water and land resources, and of ecosystems and their services, are considered alongside these drivers, as these are critical determinants of the pathways with sufficient potential to move food-producing systems towards more sustainable production. Hence, addressing the opportunities, synergies and constraints of multiple drivers will be critical for policy advice to build resilient food systems in the future.

Background

Food security, meaning access to adequate food for all, at all times, requires, inter alia, sustainable and increased production and productivity in the agricultural sectors, as well as more equitable distribution of food. In this chapter the starting point for understanding food security is grounded in the food security framework developed by FAO (EC-FAO Food Security Programme, 2008) to reflect the multifaceted risks and challenges possible along the food supply chain to attain food security. The general framework comprises four dimensions:

- *Food availability:* the availability of sufficient quantities of food of appropriate quality, supplied through domestic production or imports.
- *Food access:* access by individuals or nations to food, including access to

*E-mail: jennie.barron@sei-international.org
†E-mail: rtharme@tnc.org

resources to produce food and the ability to purchase food.

- *Food stability:* to be food secure, a population, household or individual must have access to adequate food at all times. They should not risk losing access to food as a consequence of sudden shocks (e.g. an economic, societal or climatic crisis) or cyclical events (e.g. seasonal food insecurity).
- *Food utilization:* utilization of food through appropriate diet, clean water, sanitation and health care to reach a state of nutritional well-being where all physiological needs are met.

In all these dimensions of food security, water and other ecosystem services play integral parts in both supply and impact. Hence, food security is the product of many variables, including: physical factors such as climate, soil type and water availability; the management of these factors and other natural resources (water, land, aquatic resources, trees and livestock) at the level of fields, landscapes and river basins; and losses and waste along the value chain. It also requires adequate policies and institutions in the many sectors that influence the ability of men and women to produce and purchase food, and the ability of their families to derive adequate nutrition from it. These intricate linkages mean that food security cannot be considered in isolation. The feedbacks among food production, access, reliability and utilization are essential in the context of multiple changes in society and its environment (see Box 2.1).

Drivers, which may be defined as any natural or human-induced factor that directly or indirectly causes a change in an ecosystem (Millennium Ecosystem Assessment, 2005a; Carpenter *et al.*, 2009), can be observed at global, regional and local scales, and ultimately put direct or indirect pressure on the management of natural resources. Key global drivers discussed here centre around food and water availability, because these are major influences affecting agricultural water demand and increasing the pressure on ecosystems. A workable framework of drivers and causal links affecting water stress and sustainability, as well as human well-being, is well illustrated in Cosgrove *et al.* (2012).

This chapter is focused around major drivers of change to the food security–water–ecosystems complex as loosely corresponding to those identified in the recent Foresight project 'Global Food and Farming Futures' (Beddington, 2010; Foresight, 2011); the types of drivers are similar to those of various global assessments, such as the Millennium Ecosystem Assessment (2005a,b), the World Water Assessment Programme of the United Nations (WWAP, 2009, 2012) and the Intergovernmental Panel on Climate Change (IPCC, 2012). Thus, this chapter will address the demographic drivers (i.e. population trends and changes in population preferences), the current state and trends in ecosystem services, climate change, and issues on the globalization of economies and governance.

Natural Resources and Ecosystem Health for Food

Terrestrial and aquatic ecosystems provide food for people, both as ecosytems in their natural state, for instance through forest products and inland capture fisheries, and in the form of intensively or extensively managed landscapes, such as crop and forestry systems, livestock keeping and aquaculture (see Chapter 4). Global estimates on the water needed for meeting the Millennium Development Goal (MDG) target on hunger suggest that the current appropriation of circa 7130 km^3 annually for food needs to increase to at least 12,050–13,500 km^3 by 2030 (Rockström *et al.*, 2009a). Some of this additional water may be mobilized through water savings such as improved water productivity, in particular in currently low-yielding agroecosystems (see Chapter 8).

There are fundamental differences in opportunities among, as well as within, countries, depending on their available resources of both water and investment capacity (Rockström *et al.*, 2009a). Access and control over land, water and produced capitals (e.g. financial capital, technologies) are also key factors to achieve the MDGs and increase water productivity in a way that will benefit the poor – notably women (UNEP, 2009). These different opportunities for the

Box 2.1. Hunger and food security.

The latest FAO estimates indicate that global agricultural production needs to grow by 70% between 2009 and 2050 to feed the population. The increase is due to a shift in demand towards higher value products of lower calorific content, and an increased use of crop output as feed for the rising meat demand (FAO, 2009a). At the same time, the adaptation of the agriculture sector to climate change will be a necessity for food security, poverty reduction and the maintenance of ecosystem services. In such a context, sustainable use and management of water and biodiversity resources in agroecosystems play a decisive role in providing food and income for a growing population (Nellemann *et al.*, 2009; FAO and PAR, 2011).

Despite 10 years of global commitment to reduce hunger, the number of hungry people (as measured through Millennium Development Goal (MDG) target 1A) remains more or less the same as estimated during the base year of 1990 (Fig. 2.1). Significant gains have been achieved in the past 20 years, as the relative share of hungry people has decreased from around 20% of developing country populations in 1990 to a current value of 12.5% (FAO, WFP and IFAD, 2012). Still, about 870 million people do not have sufficient food and 98% of these live in developing countries. Sixty-five per cent of the world's hungry live in India, China, the Democratic Republic of Congo, Bangladesh, Indonesia, Pakistan and Ethiopia. Women are particularly vulnerable and account for about 60% of the global hungry (FAO, 2010).

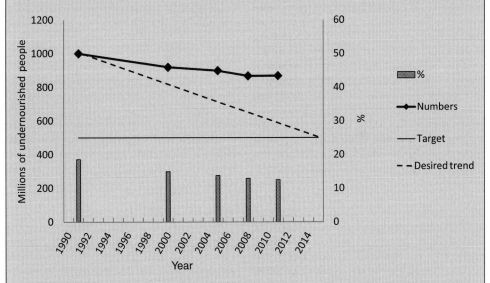

Fig. 2.1. Trends in numbers and percentages of undernourished people in the world for the period from 1990 to 2015 (last point projected), compared with the Millennium Development Goal (MDG) target of halving the number of hungry people (based on FAO, WFP and IFAD, 2012).

development of water for food security may have quite different impacts on water resource appropriation in different countries, in addition to impacts on the downstream flows that ultimately affect various water-related ecosystem services and functions. A comprehensive analysis of the need for water for food, and of the potential impacts on water-dependent ecosystem services in various landscapes, is not yet available on an aggregated global level.

Water is one of the main factors limiting future food production, particularly in the poorest areas of the world, where access to water, and its timely availability, are especially challenging. Over 1.6 billion people currently live in areas of physical water scarcity, and 1 in 10 continues to lack water for drinking and sanitation (UN, 2011). For 'business as usual' in agricultural practices, increased urbanization and changed diets, the amount of water required for agriculture to feed the world population would need to increase by 70–90% (Molden, 2007; Rockström *et al.*, 2009a). Yet humans and ecosystems already face water stress from over-abstraction and from pollution (e.g. Rijsberman, 2006; see Chapter 5). Groundwater depletion is an under-examined issue of special concern, given its critical link in sustaining irrigation and people in highly densely populated areas (e.g. Giordano and Villholth, 2007). Close to 80% of the world's population is exposed to high levels of incident threat to water availability, according to a first global synthesis that jointly considers both human water security and biodiversity perspectives (Vörösmarty *et al.*, 2010). The challenge is, therefore, to improve water productivity at the landscape or river basin level, thus accounting for a wider set of goods and services beyond agricultural produce (Ong *et al.*, 2006; Molden, 2007; see Chapter 8).

The Millennium Ecosystem Assessment sought to catalogue the state of the environment and assess the consequences of ecosystem change on human well-being, including its effects on (and the effects of) food production (Millennium Ecosystem Assessment, 2005a). It showed that the significant increases in provisioning services (largely the goods used by people) that has been achieved in recent times, in particular food production through agriculture, have, to a large extent, been achieved at the expense of reductions in other ecosystem services, such as cultural aspects or services supporting or regulating other items that people need to sustain their well-being, societies and economies. These regulating and supporting services include, among other functions, drinking water supply, flood and drought protection, nutrient recycling, regulation of pests and diseases, and the provision of habitats for flora and fauna (for more on ecosystem services see Chapter 3).

The rural poor and marginal groups continue to have direct reliance on the ecosystem services of healthy natural ecosystems. In times of natural or anthropogenic shocks, such as droughts, floods, fires or market price volatility, there are few, if any, safety nets for ensuring that even their most basic nutritional needs are met. These groups of people also have less capacity to cope with the situation, or to find

substitutes, when ecosystems and their services begin to degrade, and therefore are increasingly and more immediately vulnerable to such degradation (WRI, 2005).

Ecosystem deterioration, and the resultant loss of integrity, biodiversity and valued ecosystem services, along with the risk of reduced system resiliency to future shocks, must be more adequately factored into our understanding of drivers and the complex system feedbacks that their trends induce to safeguard food security in the future (Keys *et al.*, 2012). Environmental degradation generates multiple negative feedbacks on food production systems, and on the livelihoods and human well-being they support. Depleted, fragmented and polluted river systems, lakes and aquifers already bear testament to these interrelationships. For instance, some 65% of global river discharge, and the aquatic habitat that water supports, are under moderate to high threat (Nilsson *et al.*, 2005; Dudgeon *et al.*, 2006). Such documented alterations to ecosystem health expose the currently untenable situation of accelerated degradation of natural and agroecosystems, especially wetlands (Millennium Ecosystem Assessment, 2005b), and the resultant declines in and unintended consequences for human ecosystem benefits (for further discussion pertaining to wetlands see Chapter 7).

Biodiversity is a central indicator for the state of the global environment and ecosystem services (see also Chapter 9). It has been suggested that the current rates of species extinction are far beyond what is considered a 'safe operating space for humanity' (Steffen *et al.*, 2011). Indeed, an assessment of 31 different indicators of the status of global biodiversity in relation to the Convention on Biological Diversity (CBD; initiated in 1992) target of achieving a significant reduction in the rate of biodiversity loss by 2010 was unequivocal in demonstrating that the rate of biodiversity loss is not lessening at a global scale (Butchart *et al.*, 2010). In this study, state-of-biodiversity indicators pointed to declines in biodiversity without a significant reduction in its rate of decline (Fig. 2.2, dotted line 'State'). This was coupled with an acceleration in the risk of species' extinction, with only freshwater quality and trophic integrity in the marine ecosystem showing marginal improvement. In direct contrast, various indicators of the pressures (or indirect drivers) on our ecological assets, such as the ecological footprint, which reflects aggregate resource consumption, nitrogen pollution and climatic impact, showed increases (Fig. 2.2, solid line 'Pressure'). Practice and policy responses (among these, the extent of protected areas and official development assistance for biodiversity), while encouraging in their increases and, in a few cases, in their local success, presently remain inadequate to check the trend of deterioration (Fig. 2.2, dashed line 'Response'). Perhaps unsurprisingly in this context, though based on a poor information base, the benefits that humans have derived from their natural capital were also found to be in accelerated decline; this is perhaps most significant for the more than 100 million poor people inhabiting remote

areas within threatened ecosystems (Butchart *et al.*, 2010) who are likely to be particularly dependent upon the ecosystem services of healthy ecosystems with high biodiversity. There is an urgent need to identify new and improved local and global governance models that can ensure sustainable food production, while managing ecosystem services and bio-diversity in synergy.

Alongside water resources, the present state of land, soils and their biodiversity may present the fundamental challenge for the future of food security (Bossio and Geheb, 2008), with some 11.7% of global land cover already converted to cropland (for which Steffen *et al.* (2011) propose a planetary boundary of 15%). Moreover, a recent report by FAO (2011) entitled *The State of the World's Land and Water Resources for Food and Agriculture (SOLAW)* concluded that growth in food production must take place on existing land. That is, current low-producing agricultural land

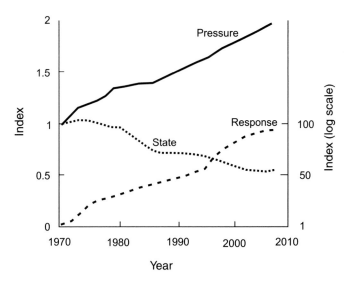

Fig. 2.2. Aggregated indices of the state of and pressure on biodiversity (left-hand y-axis), and the responses of biodiversity to protection, policy and aid measures (right-hand y-axis) over the period 1970–2010. The state of biodiversity (dotted line 'State') is based on nine indicators that cover species' population trends, habitat extent and condition, and community composition; pressure on biodiversity (solid line 'Pressure') is based on five indicators of ecological footprint, nitrogen deposition, numbers of alien species, over-exploitation and climatic impacts. The response (dashed line 'Response') of biodiversity to various measures is based on six indicators that cover protected area extent and biodiversity coverage, policy responses to invasive alien species, sustainable forest management and biodiversity-related aid (after Butchart *et al.* 2010). Values in 1970 were set to 1 for 'State' and 'Pressure', and to 0 for 'Response'.

will need substantial investments to become productive as well as to avoid taking new land under cultivation. According to the *SOLAW* report, more than one third of agricultural land is already severely or moderately affected by land degradation. Moreover, there is a mismatch between resource availability for increasing production, i.e. access to relatively arable land and reasonable quality water resources, and expected needs from the places where food-insecure and poverty-affected people live and will live in the near future. This outset provides a fundamental challenge on how to ensure food security, because the current state of resources is already degraded; particular regions at risk for soil and water resources have been identified in the highlands of East Africa, and in South and East Asia. Under current agricultural practices, this would result in an increasing demand for land of up to an additional 200 million ha by 2030 (Bindraban *et al.*, 2010) for food and feed only. This does not even consider the potential impact of people's needs for fibre, timber and fuel, which also require land.

Demographic and Social Drivers

Understanding trends in population size and associated demographics will be critical to estimating the future demand for food. A review of how reliable population projections are showed that by 2050 there will be between 8 and 10 million people, with most growth in developing countries (Lutz and Samir, 2010). Hence, there are two aspects to the driver relating to food security and demographic change at the global scale[1]. First, in order to feed approximately 9 billion people by 2050, food production has to increase (probably double, according to Molden, 2007). Secondly, as the global population increases its wealth, in terms of more income per capita, food composition will increase and change (Fig. 2.3). Higher incomes result in choices of food that appropriate more water per produced energy unit (Fig. 2.4; Lundqvist, 2006), although this depends on whether the diet is vegetarian or mixed. The change of water appropriation for various diets is well

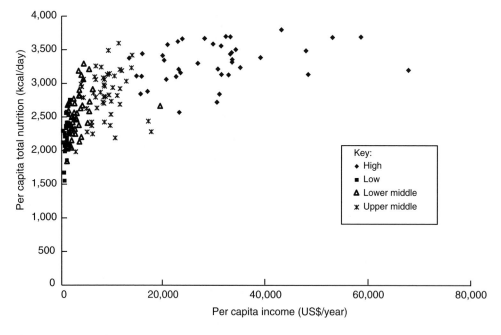

Fig. 2.3. Per capita dietary consumption (kcal/day) versus per capita income for various countries according to the World Bank classification (based on data from FAOSTATS, 2012 and World Bank, 2012).

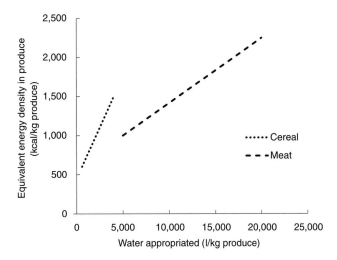

Fig. 2.4. Principal relations between water appropriated for cereal and meat energy content in food (adapted from Lundqvist *et al.*, 2007).

established (e.g. Molden, 2007; Hoekstra and Mekonnen, 2012). As increased energy is derived from animal protein, the amount of water needed to produce that energy increases. In terms of grain equivalents (GE), daily consumption generally varies from 1 to 1.5 kg GE/person for a vegetarian diet (using 1000– 1500 l water) and 4 to 5 kg GE/person in wealthy societies (meat rich diet; using 4000– 5000 l). Demand for aquaculture products such as fish and shrimp also continues to rise, which means further demand for freshwater resources (Bostock *et al.*, 2010; Hoanh *et al.*, 2010; FAO, 2012b). Thus, more water will be needed as populations increase wealth and consume more animal protein (Fig. 2.4). Near future changes in income for large populations in emerging (upper and lower middle income and low income) countries will have substantial impacts on the current demand for food production and food security, and the water used to produce this food (Box 2.2; see also Tilman *et al.*, 2011).

Parallel to the specific drivers of livestock production, drivers for fisheries that further push these into aquaculture have been identified by Bunting (2013). In addition to threats to freshwater habitat, there are drivers on the demand side in access to resources and

in risks margins for the people whose livelihoods depend on fish. Integrated approaches at various levels are required to sustain critical ecosystem services that support fish production (UNEP, 2010).

One of the traditional and adaptive responses to environmental stress has been human migration, often undertaken in an attempt to diversify sources of income, another important demographic driver. While earlier reports suggested that climate change would be a main driver of migration, in reality, socio-economic circumstances are the key determinants (Tacoli, 2011). However, it is clear that most migration takes place south to south, rather than south to north (Tacoli, 2011). Thus, the countries and locations currently dealing with immigration and new settlements are areas that are already pressured to attain food security (Sharma, 2012).

In 2008, the world's population was split evenly between urban and rural dwellers. By 2030, there will be 1.8 billion more urban dwellers and 100 million fewer rural inhabitants (WWAP, 2009). Urbanization, projected to continue at an accelerating pace, is expected to account for 70% of the world population in 2050. As people move to cities and alter their lifestyles, urban upper and middle classes

Box 2.2. Focus: drivers of livestock systems.

Livestock production is the single largest land user globally, with grassland covering 25% of the earth's land area, and land dedicated to feed crops making up one third of the global cropped area (Herrero *et al.*, 2010). In developing countries, however, livestock feed is mainly derived from crop residues and from rangeland with low potential for cropping. Livestock production contributes 53% and 33% of the agricultural gross production in industrial and developing countries, respectively. Developing countries produce 50% of the beef, 41% of the milk, 72% of the lamb, 59% of the pork and 53% of the poultry globally. Livestock is an integral part of mixed crop–livestock systems, and these produce close to 50% of the global cereals. The importance of the livestock sector is also clear from the value of production, as milk has the highest value of production of all commodities globally, followed by rice (second) and by meat from cattle, pigs and poultry (third) (Herrero *et al.*, 2010).

Many animal food products from livestock and poultry will depend on grain as the limits to production on grazing land are reached (Peden *et al.*, 2007). Moreover, growth in the industrial pig and poultry sectors in South America and Asia will create the need for additional grain for feed: by 2050, more than 40% of global cereal use will be for feed purposes (Herrero *et al.*, 2009). Because rich countries already consume high amounts of livestock products, the growth in demand is predominantly a developing country phenomenon (Table 2.1), where approximately a billion poor people are supported by livestock.

Table 2.1. Current and projected consumption of animal products (from Herrero *et al.*, 2009).

Countries	Year	Annual per capita consumption		Total consumption	
		Meat (kg)	Milk (kg)	Meat (Mt)	Milk (Mt)
Developing	2002	28	44	137	222
	2050	44	78	326	585
Developed	2002	78	202	102	265
	2050	94	216	126	295

For poor smallholder farmers, livestock provide diverse products and services (e.g. they represent a major source of draught power) and an insurance against various shocks. Livestock are also an income source, and they provide livelihood diversification and improved nutrition. In addition to urbanization and changes in diet, other drivers also affect livestock production and illustrate how food security and consumption may drive agriculture and influence the management of agroecosystems (Table 2.2).

Table 2.2. Balancing food production, maintenance of ecosystem services and poverty reduction in livestock systems of the developing world through policy, investment and technology (adapted from Herrero *et al.*, 2009, 2010).

Drivers and pressures	Policy needs	Investment needs	Technology needs
Agropastoral systems			
Significant rural–urban migrations, more conflicts, higher numbers of vulnerable people, increases in livestock numbers in some places, significant impacts of climate change in places, resource degradation	Frameworks for diversifying income sources, including payments for ecosystem services and others, insurance-based schemes	Roads, livestock markets, health and education establishments, development of water sources, food storage systems, telecommunications	Matching livestock breeds to the agroecosystems, livestock species changes in some places, suitable crops if required, early warning systems, mobile phone based telecommunication products, prices information and others

Drivers and pressures	Policy needs	Investment needs	Technology needs
Table 2.2. Continued			

Drivers and pressures	Policy needs	Investment needs	Technology needs
Extensive crop–livestock systems			
Manageable increases in population density but significant rural–urban migrations, potential for increased crop and livestock production through intensification and though large impacts of climate change in some places	Policies to create incentives and an enabling environment to produce food in these regions, appropriate credit, land tenure rights, incentives for public–private partnerships, service and support institutions	Infrastructure: roads, postharvest storage systems, water sources and storage, health and education establishments, markets, development of value chains, involvement of the private sector, product processing plants, telecommunications	Crop varieties suitable for the agroecosystem, fertilizers and agricultural inputs, livestock feeds, breeding systems, livestock vaccines and health management
Intensive crop–livestock systems			
Large increased population densities, reductions in the primary productivity of crops, water scarcity or soil fertility constraints, large increases in livestock numbers, increases in food prices, potential food insecurity, environmental degradation, increases in zoonotic and emerging diseases	Regulations for intensification/ de-intensification, monitoring and evaluation frameworks for assessing environmental impacts, appropriate regulatory frameworks for global food trade	Infrastructure to support value chains – ports, railways, cold chains, processing plants, supermarkets and storage facilities; human capacity development to improve management skills	Options with high efficiency gains: more crop per drop, more crop per unit of fertilizer, species or animals with improved conversion efficiencies of feed into milk and meat
Industrial landless systems			
Most growth in monogastric production, heavy dependence on grains as feed, expansion into areas further away from centres of demand as transport efficiency develops	Regulations for intensification/de- intensification, monitoring and evaluation frameworks for assessing environmental impacts; appropriate regulatory frameworks for global food trade	Infrastructure to sup- port value chains – ports, railways, cold chains, processing plants, supermarkets and storage facilities	Animals with improved conversion efficiencies of feed into milk and meat, more efficient diet for- mulation, technologies for waste disposal

consume more energy and water-intensive diets (Kearney, 2010). Wealthier urban inhabitants are likely to consume both more calories and have higher protein diets (especially processed foods, and dairy and meat products, which have higher water requirements per calorie) than their rural counterparts (von Braun, 2007; Cirera and Masset, 2010; de Fraiture and Wichelns, 2010; Fig. 2.4).

Since the year 2000, a particular change related to demography is the increasing demand for energy from renewable resources (see also Box 8.3 in Chapter 8). The production of biofuels, particularly ethanol and biodiesel for use in the transport sector, has tripled and is projected to double again within the next decade (FAO, 2008b). This increase has been driven largely by policy support measures in the developed countries that are seeking to mitigate climate change, enhance energy security and support the agricultural sector. If the world switches predominantly from fossil fuels to the production of biofuels, this will have immense impacts on ecosystems and water availability (de Fraiture *et al.*, 2008; Bindraban *et al.*, 2009). Currently, biofuels account for 0.2% of total global energy consumption, 1.5% of total road transport fuels, 2% of global cropland, 7% of global coarse grain use and 9% of global vegetable oil use (FAO, 2008a). These shares are projected to rise over the next decade, as patterns of energy consumption shift in rural and urban areas; at present, two thirds of the world's poorest people still rely on fuelwood and charcoal as their major source of energy for heat and cooking (which represents over 40% of the wood removal from forest globally; FAO, 2006).

Climate Change

Future food, fodder and fibre production and ecosystem services will be under additional risk and uncertainty from climate change. Fundamental 'climate-related tipping points' have been proposed, which may seriously affect food security in various regions currently struggling with food security and poverty, including West Africa and South Asia (Lenton *et al.*, 2008), as well as from an increase in extreme events such as droughts and floods (IPCC, 2012). Recent studies of temperature trends confirm that warming is happening faster than anticipated and at a global scale, with extreme temperature events no longer being extreme as they occur more often (e.g. Hansen *et al.*, 2012).

Predicting the effects of global climate change is a process that is daunting in scale and uncertain at best in its application. Some ecosystems are more vulnerable to the negative effects of climate changes than others, with freshwater systems identified as being particularly vulnerable (Bates *et al.*, 2008). In certain cases, their resilience may be undermined to the extent that irreversible losses or complex shifts may occur in biodiversity and in various ecosystem services, such as the regulation of pests and water flows (Fischlin *et*

al., 2007; UNEP, 2007). Climate change is predicted to affect agriculture and forestry systems through higher temperatures, elevated carbon dioxide (CO_2) concentration, changes in precipitation and the pattern and timing of runoff, and increased pressure from weeds, pests and diseases (FAO, 2009b; Le Quesne *et al.*, 2010).

Of particular concern are the potential impacts on freshwater resources, as rainfall (or indeed snowmelt) patterns change because alterations in rainfall distribution, combined with decreases in volume, can result in significant decreases in streamflow. There are also suggestions of 'tipping point' features in hydrological systems, in which a small change potentially results in large impacts. A study of basins on the African continent modelled climate change as a reduction of 10% in annual rainfall. This might potentially result in a 25–75% decrease in streamflow in the 400–800 mm rainfall zones (de Wit and Stanckiewicz, 2006), i.e. a 'tipping point' feature in the response of streamflow with a marginal reduction of rainfall. The study also indicated a greater sensitivity of surface water availability in regions already subject to high seasonal and inter-annual rainfall and surface water availability, which applied to agriculture, society and ecosystem services. Other important features of the modelled climate change included the timing of the onset of rainy seasons, where new evidence is emerging that these – in, for example, the Sudano–Sahelian zone – are becoming less distinct with more 'false onsets' (de Wit and Stanckiewicz, 2006). Similar trends have been identified for the onset of the South Asian monsoon (e.g. Asfaq *et al.*, 2009; Washington *et al.*, 2012).

As agriculture is particularly dependent on the hydrological cycle, food production will obviously be greatly affected by changes in precipitation, streamflow, soil moisture and evapotranspiration. Local agricultural production may increase or decrease under conditions of climate change (and agriculture itself has well-established positive and negative feedbacks to climate change, see Box 2.3). Uncertainty is high for projections of rainfall patterns, and, as a result, the impact on major crop yields has been shown to vary significantly for different regions and scenarios of climate

---

Box 2.3. Agriculture-driven feedbacks on climate change.

Climate change is clearly a driver that will affect food and water security for the foreseeable future, albeit with a high degree of uncertainty in the precise way in which the impact will be felt for specific locations and crop and crop–livestock systems. As knowledge of its impacts increases, so should understanding also improve of the diversity and complexity of the concomitant feedback effects from agricultural food production on climate change.

For example, by recent estimates, the agricultural sector as a whole accounts for roughly 14% of global greenhouse gas (GHG) emissions, of which three quarters comes from developing countries (Parry *et al.*, 2007; FAO, 2009b). The contribution of livestock (especially cattle) production to global anthropogenic GHG emissions alone has been estimated at 18%, through methane (CH_4, 25–30%), carbon dioxide (CO_2, 30%) and nitrous oxide (N_2O, 25–30%) (Steinfeld *et al.*, 2006; O'Mara, 2011); these amount to more emissions per kilocalorie when compared with crops (for more details on emissions from livestock production systems see, e.g. Tilman *et al.*, 2001; Pelletier and Tyedmers, 2010; Bouwman *et al.*, 2011). Emissions vary both regionally and in intensity, mainly in relation to the species (monogastrics are more efficient than ruminants), the product (milk, white meats and eggs are more GHG efficient than red meat) and the productivity of the animal (the higher the productivity the lower the emissions per unit of product; see FAO, 2010). In turn, these aspects depend on feed type, quantity, quality and provenance, and on the manure management system implemented. Stored manure and wet rice cultivation also contribute CH_4 to the atmosphere (Mosier *et al.*, 1998), while excessive and inappropriate fertilizer applications result in N_2O emission (Smith and Conen, 2004; Oenema *et al.*, 2005), and CO_2 is released from microbial decay or the burning of plant and soil organic matter (Janzen, 2004).

Conversely, many agricultural and natural ecosystems serve as carbon sinks, absorbing atmospheric CO_2 and thereby potentially slowing down climate change. Overall, terrestrial ecosystems have taken up approximately 25% of anthropogenic carbon in the past century (WWAP, 2009); however, ecosystem degradation is known to be limiting such buffering capacity. For example, the world's grazing lands store 10–30% of total soil carbon (Schuman *et al.*, 2002). Sahelian rangelands are highly degraded, but with proper management they could potentially capture 0.77 t carbon/ha annually (Woomer *et al.*, 2004; see also Chapter 4). There is also increasing evidence for other feedback linkages between factors such as changes in land use and land cover, and their impacts on precipitation (e.g. Gordon *et al.*, 2010), for example, through reduction in tree cover (Makarieva *et al.*, 2010).

In addition to experience of the effects of such positively and negatively reinforcing feedback loops on climate change as a driver, there is, of course, considerable knowledge of best practices for mitigating climate effects (e.g. Metz *et al.*, 2007), with up to 70% of the potential for technical and economic mitigation coming from agriculture in developing countries (FAO, 2009b).

---

change (e.g. Lobell *et al.*, 2008; Knox *et al.*, 2011). Several projected trends will adversely affect food security in developing countries, particularly in Africa, and increase the dependency of many of these countries on food imports. It is estimated that climate change will reduce Africa's potential agricultural output by 15–30% by the 2080–2100 period (FAO, 2009b; Ericksen *et al.*, 2011).

Climate change will also have a variety of effects on the water sector itself, including effects on its institutions and their inherent capability for successful adaptation (Cook *et al.*, 2010). Water planners will be less able to use historical data to plan, design or operate hydrological systems; though new prediction models are under development, which will facilitate the necessary policy solutions

(Molden, 2007). However, the current trend in reduced hydro-meteorological monitoring (e.g. synoptic weather stations, streamflow gauging stations) does have an impact on the availability of monitored data to ground-truth models, in addition to its effect on the generation of statistical trends of change, such as in rainfall amounts and distribution (e.g. Hannerz, 2008). With increasing variability in rainfall (amounts and events) it will be more important to store water in the soil (as soil moisture) and in the landscape (as ponds and dams) at various scales, to reduce the risk of additional crop and livestock losses through climatic extremes (Bates *et al.*, 2008; McCartney and Smakhtin, 2010). As an adaptation strategy, increasing the storage of water to bridge dry spells, droughts and dry seasons may need careful consideration to maximize synergies between multiple uses of water in landscapes, such as the use of water by agriculture and ecosystems within and downstream from water storage interventions (e.g. environmental flows, see Chapter 10).

In contrast, the climate change impact on temperature is more consistently modelled in climate change scenarios. It is increasingly a concern that 'worst case scenarios' appear to be confirmed by measured global temperatures during the last decennium. Although the increase in the average temperature may benefit some areas of the globe, it is likely to have a negative effect on yields in current crop-producing areas, such as southern Africa, central Asia and Brazil (Lobell *et al.*, 2008); a higher degree of uncertainty remains for some areas. Various crops are also significantly differentially sensitive to temperature, as well as to the joint change in climate brought about by the combination of temperature increases and altered rainfall patterns (Parry *et al.*, 2007). Current outlooks for climate change suggest that it will disproportionately adversely affect sub-Saharan Africa (Ericksen *et al.*, 2011), where food production per capita is already the lowest globally (McIntyre *et al.*, 2008), and lack of food security and accessibility are recurrent problems at local and regional levels. The adequacy of forecasts is further complicated by the impacts that agriculture itself may have on climate change (Box 2.3).

Globalization of Economies and Governance

A third driver of significance for the linkages between food security, water and ecosystem services is the role that global and local markets, and also the governance of resources access and use, may play in the future. There are currently a number of economic, market-related issues that are affecting, and may in the near future have further significant impacts on, food and water security.

As a driver, global food commodity prices play an important role as producer incentives. While up to 80% of the produce of smallholder farmers is sold at local markets, these markets are not disconnected from global markets and prices. Therefore, as consumers, smallholder farmers and rural populations in developing countries are affected by price hikes, without necessarily being able to benefit from them as producers. The 2007/8 and 2010/11 world-wide price hikes on staple foods (e.g. FAO, 2012a) are examples that show how food security is affected by global drivers at multiple scales. The most recent rise in prices has driven 110 million more people into poverty, both in rural and urban areas. Over the next decades, food prices are predicted to remain at current levels (OECD and FAO, 2012; Fig. 2.5). The sudden increase in food prices that 2006/7 brought was largely unanticipated and has resulted in an increased burden on the

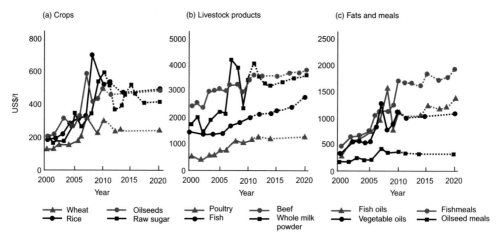

Fig. 2.5. Past actual and projected price development at global markets for (a) key crops, (b) livestock products and (c) fats and meals, for the period 2000–2020 (OECD and FAO, 2011).

poor, who already spend one half to three quarters of their income on food. Major food-producing countries have restricted exports of food to keep costs down domestically, which has raised international food prices even more. Increased food costs are likely to push governments to invest more in agricultural productivity, but this will take years to offset the current high food prices (WWAP, 2009).

There are multiple reasons for these price spikes, which are only partly explained by the agronomic conditions of food production. Increasingly, food is traded as a commodity, and thus is subject, for example, to similar financial speculations such as those for housing, metals and insurances. Some advocate that the global food commodity market is non-transparent and hence inherently flawed as market mechanisms cannot operate (e.g. Oxfam, 2011). A recent review by Huchet-Bourdon (2011) on agricultural commodities and global price volatility over the last 50 years suggests that global markets are being increasingly interconnected. Consequently, price volatility characteristics in the past cannot readily be compared with today's market conditions, where price information and commodities are being shifted much faster.

In the developing world, more than 1 billion people still rely on their own production of food for food security (IFAD, 2010), and approximately 450 million are actively engaged in farming as either self-employed or employed. On a global level, the number of people directly relying on agriculture has increased marginally from 2.2 billion in 1980 to approximately 2.6 billion today. This growth of 20% from 1980 is substantial in absolute numbers but is still far less than the 90% increase in the corresponding non-agricultural share of the population during the same time period (Fig 2.6). As a driver of change, it will be important to consider the implications of this shift, for instance with regards to local–regional availability of labour and the skill sets needed to ensure the transformation of food systems to more desirable, sustainable states in the long term.

Still, the farming community that is producing crops to ensure food security for themselves and other consumers is by far more diverse and multifaceted than are the global retailers that are transferring produce and food commodities between producers, retailers and consumers. On disaggregating cereal exporters globally, for example, it emerges that a handful of nations supply 60–80% of globally traded cereals. Similar statistics can be found for other key agricultural commodities, such as soybeans (or products thereof), cocoa, sugar, wine, and fibres such as cotton. Thus, a small group of countries

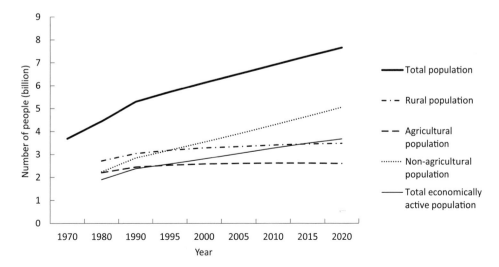

Fig. 2.6. Actual and projected total, rural and agricultural population of the world (after FAOSTAT, 2012).

constitutes a major player in food production and food security at key times and locations.

In a similar fashion, another driver of change that may be central for food security is the concentration of trade in food and agricultural commodities within a limited set of multinational corporations, traders and retailers. According to Oxfam (2011), only three major companies control 90% of global cereal trade. Yet a recent analysis by OECD and FAO (2011) of major food commodities[2] indicated that market thinness (i.e. measures of the number of actors trading) is not increasing but slightly decreasing both on the export (supply) and import (demand) sides (Liapis, 2012). The study, which used three measures of market thinness, revealed that the number of actors on the export side is in the order of 10–100, whereas major importers are in the order of 100–200. So there is a significant step up in magnitude to reach both the numbers of primary producers (2.6 billion as above) and consumers (around 7 billion globally). This concentration of trade and markets can have a significant impact on food security. Despite the study on market actors and market thinness by Liapis (2012), it is clear that the power of the markets of the major food commodities is a challenge if it is taking place within non-transparent fora. First, there is little record of the trading, volume, value of trade or actual registration of the private companies concerned. Secondly, a concentration of trade may affect the way that food is produced, including standards and quality, as well as potentially affecting choices of production systems. Both farmers and consumers may in the end be affected by this concentration in the trade and retailing of food commodities.

Governance from global to local scales is important to set the vision and pathway for the integrated, cross-sectorial management of water, food security and ecosystem services.[3] The current state of food trade is globally complex, with private and public interventions, national and international rules, regulations and subsidies affecting agriculture, food production systems and trade. While arguably as complex, presently there is a more coherent consensus on the governance of water resources (including the wide adoption of integrated water resources management, or IWRM; see Chapter 10).

As a global driver, governance principles are being put in place for sustainable water allocation for food production and security, in particular at national level and at the transboundary basin scale. These governance principles operate within the same space as the negotiation and accounting of other societal and infrastructure demands on the same water resources. However, there is scope for further development, in particular to account for the water needs of ecosystem services (see Chapter 10). Moreover, explicit accounting for water demands as ecosystem services is not necessarily better in countries with high development indices (Harlin, 2011). As a driver of change, the governance of water, and of land and biodiversity, will need to be taken into account in forecasting food security at local, regional or global scales. In the case of the coupled natural–human systems that are important for food production, not only do the types of social, economic and political settings (e.g. economic development, demographic trends, political stability, government resource policies, market incentives, media organization) set the stage for sustainability, but the system of governance is itself a subsystem central to the whole (Ostrom, 2009). Evidence across multiple cases suggests that there are conditions where resource users have self-organized to manage and improve resource governance towards more sustainable pathways (Ostrom, 2009). These examples can be used to inform other cases of less successful governance and development.

There is a range of sources of funding for developing food security and food production systems in currently low-producing and poverty-affected regions. Important global and local drivers of investments may be public, private or external North–South overseas development aid (ODA). Development aid to agriculture decreased by some 50% between 1980 and 2005, to an approximate 7.2 billion US$/year for bilateral and multilateral funds, even though total official development assistance increased significantly by 112% over the same period (OECD-DAC, 2010, 2011; Lowder and Carisma, 2011). This meant that the share of aid funds going to the agricultural sector fell from 17% in 1980 to 3.8% in 2006, with the same downward trend observed in national budgets. At the same time, the global commitment to address food security (MDG Target 1A) by halving the amount of hunger by 2015 is a powerful vision that still guides millions of US dollars of ODA. This reduction of North–South transfers has given an opportunity to new actors and policies of change, for example in Africa. At a continental level, the African Union commitment is to devote 10% of national gross domestic product (GDP) towards agricultural sector development in order to address food security in its respective countries (African Union, 2003). Another large driver of investment is the transfer of individual remittances, from north to south or from urban to rural. The amounts are globally of a similar order of magnitude to the total annual ODA, but there is limited synthesized knowledge on how reinvestments are made on the receiving end. Further knowledge of the source and use of investments is needed to determine how they are currently affecting food security and its linkages with water and ecosystem services.

Conclusions

The future of food security, and with it water resources and ecosystem services, is affected by a range of external drivers of change at global and local scales, often with uncertain outcomes. In this chapter, key drivers have been discussed that have potential multiple, sometimes coupled, impacts – namely, demographic change, climate change, and economic markets and governance. The purpose is to ensure that we address food security–water–ecosystem service issues in multidimensional and interconnected ways in global and local systems, as they are affected by and have impacts upon a range of drivers important for human well-being. That fundamental thresholds of the earth's biochemical cycles have been exceeded (Rockström *et al.*, 2009b; Barnosky *et al.*, 2012) suggests that ecosystems and ecosystem services are already in precarious states and potentially subject to undesirable tipping points. Humanity's demand for increased food,

fodder and fibre is on a trajectory towards fundamental detrimental impacts on ecosystem services, at various scales, unless immediate action to reinvent and more responsibly manage our food production system is taken.

As Butchart *et al.* (2010) and others have attested, efforts to date to slow down the loss in natural capital that encompasses the biodiversity (from habitats and species to genetic diversity) and various ecosystem services that are so valuable for food security have been grossly insufficient. Moreover, Butchart *et al.* (2010) underscored the 'growing mismatch between increasing pressures and slowing responses' (Fig. 2.2).

It is all too evident that agricultural practices need to become more deliberately systemic, creating synergies between production systems and ecosystem health, and ensuring productive and resilient landscapes for multiple benefits (Molden, 2007; Gordon *et al.*, 2010; Turral *et al.*, 2010). Appropriate strategies, safeguards, options and technical solutions need to be developed and applied to ensure that water can provide for a wide set of ecosystem services, including agriculture, and for diversified incomes and food security in an environmentally sustainable manner. These approaches should be based upon a better understanding of the values and benefits, as well as the functioning, of ecosystems – be they terrestrial, aquatic or marine – and also of their interrelations with the quantity and quality of water.

The pressure of consumption and demand by the processing industry for certain characteristics of produce can be a major impact on production. There is great need for additional knowledge on how these drivers change agricultural production systems, and what the consequences are for water and ecosystem services at local and at aggregated global levels. The knowledge and skills to achieve change will be critical at farm and at management levels as well, so as to improve food production systems. Ultimately, multiple drivers will need to be explored in combination to identify and best characterize more sustainable agricultural productions systems. Such efforts are urgently needed to find synergistic pathways of development for addressing food security and sustainable water and ecosystems management. The future research and management of agriculture for ecosystem services and water must consider a range of drivers of change, with high or low degrees of certainty, in order to support best-bet investments and policy action.

Acknowledgements

We sincerely appreciate contributions from Petina L. Pert and Renate Fleiner, and also from Luke Sanford, Nishadi Eriyagama, Tilahun Amede, Catherine Muthuri, Fergus Sinclair and Katrien Descheemaeker, who provided an earlier background setting to this chapter.

Notes

[1] There are other issues relating to population and food systems, in particular issues of over-consumption and obesity, as well as changing age distribution in populations. Although these are important drivers for food systems, they are not considered here, as the scope of the chapter is on food production and security related to ecosystems and water resources.

[2] The study included maize, rice, wheat, sugar (raw, refined), beef, butter, soy (bean, oil) and milk (cream, powder).

[3] A full treatment of the topic of governance as it pertains to environment and energy is beyond the scope of this book. Other factors also affect agricultural production systems and ecosystems but have not been thoroughly discussed in this chapter in order to maintain the focus on linkages with water over the entire volume. These include changes in global governance of key resources, energy price development, and advances in technologies, in production, in processing and in consumption, as well as in information technologies. The role and value of innovation in contributing to more efficient and sustainable production systems have not been addressed; nor has the value of research and the effect of improved governance in contributing to the understanding of water–food–ecosystem complexities.

References

African Union (2003) Assembly/AU/Decl.7 (II) Declaration on agriculture and food security in Africa. In: *Assembly of the African Union, Second Ordinary Session 10–12 July 2003, Maputo, Mozambique.* African Union, Addis Ababa, pp. 10–11. Available at: www.nepad.org/system/files/Maputo%20 Declaration.pdf (accessed December 2012).

Asfaq, M., Shi, Y., Tung, W.W., Trapp, R.J., Gao, X., Pal, J.S. and Diffenbaugh, N.S. (2009) Suppression of South Asian summer monsoon precipitation in the 21st century. *Geophysical Research Letters* 36: L01704, doi:10.1029/2008GL036500

Barnosky, A.D. *et al.* (2012) Approaching a state shift in earth's biosphere. *Nature* 486, 52–58. doi:10.1038/nature11018

Bates, B.C., Kundzewicz, Z.W., Wu, S. and Palutikof, J.P. (eds) (2008) *Climate Change and Water.* Technical Paper of the Intergovernmental Panel on Climate Change, IPCC Secretariat, Geneva, Switzerland. Available at: www.ipcc.ch/pdf/technical-papers/climate-change-water-en.pdf (accessed May 2011).

Beddington, J. (2010) Preface: Global food and farming futures. In: Godfray, H.C.J., Beddington, J.R., Crute, I.R., Haddad, L., Lawrence, D., Muir, J.F., Pretty, J., Robinson, S. and Toulmin, C. (eds) *Food Security: Feeding the World in 2050. Papers of a Theme Issue. Philosophical Transactions of the Royal Society B* 365, 2767. doi:10.1098/rstb.2010.0181

Bindraban, P.S., Bulte, E. and Conijn, S. (2009) Can biofuels be sustainable by 2020? *Agricultural Systems* 101, 197–199. doi:10.1016/j.agsy.2009.06.005

Bindraban, P.S., Jongschaap, R.E.E. and van Keulen, H. (2010). Increasing the efficiency of water use in crop production. In: Sonesson, U., Berlin, J. and Ziegler, F. (eds) *Environmental Assessment and Management in the Food Industry: Life Cycle Assessment and Related Approaches.* Woodhead Publishing Series in Food Science, Technology and Nutrition No. 194, Woodhead Publishing, Cambridge, UK, pp. 16–34. Available at: www.woodheadpublishing.com/en/book.aspx?bookID=1564 (accessed May 2012).

Bossio, D. and Geheb, K. (eds) (2008) *Conserving Land, Protecting Water.* Comprehensive Assessment of Water Management in Agriculture Series 6. CAB International, Wallingford, UK, in association with

CGIAR Challenge Program on Water and Food, Colombo and International Water Management Institute (IWMI), Colombo.

Bostock, J., McAndrew, B., Richards, R., Jauncey, K., Telfer, T., Lorenzen, K., Little, D., Ross, L., Handisyde, N., Gatward, I. and Corner, R. (2010) Aquaculture: global status and trends. *Philosophical Transactions of the Royal Society B* 365, 2897–2912. doi:10.1098/rstb.2010.0170

Bouwman, L., Goldewijk, K.K., Van Der Hoek, K.W., Beusen, A.H.W., Van Vuuren, D.P., Willems, J., Rufino, M.C. and Stehfest, E. (2011) Exploring global changes in nitrogen and phosphorus cycles in agriculture induced by livestock production over the 1900–2050 period. *Proceedings of the National Academy of Sciences of the United States of America.* doi 10.1073/pnas.1012878108; correction (2012) doi:10.1073/pnas.1206191109.

Bunting, S.W. (2013) *Principles of Sustainable Aquaculture: Promoting Social, Economic and Environmental Resilience.* Earthscan, London, from Routledge, Oxford, UK.

Butchart, S.H.M. *et al.* (2010) Global biodiversity: indicators of recent declines. *Science* 328, 1164–1168. doi:10.1126/science.1187512

Carpenter, S.R. *et al.* (2009) Science for managing ecosystem services: beyond the Millennium Ecosystem Assessment. *Proceedings of the National Academy of Sciences of the United States of America* 106, 1305–1312. doi:10.1073/pnas.0808772106

Cirera, X. and Masset, E. (2010) Income distribution trends and future food demand. *Philosophical Transactions of the Royal Society B* 365, 2821–2834. doi:10.1098/rstb.2010.0164

Cook, J., Freeman, S., Levine, E. and Hill, M. (2010) *Shifting Course: Climate Adaptation for Water Management Institutions.* World Wildlife Fund (WWF-US), Washington, DC. Available at http://www.adaptiveinstitutions.org/Shifting_Course.pdf (accessed February 2013).

Cosgrove, C., Cosgrove, W.J., Hassan, E. and Talafré, J. (2012) Chapter 9. Understanding uncertainty and risks associated with key drivers. In: WWAP (World Water Assessment Programme) (2012) *The United Nations World Water Development Report 4 (WWDR4): Managing Water Under Uncertainty and Risk (Vol. 1).* United Nations Educational, Scientific and Cultural Organization (UNESCO), Paris, pp. 259–275.

de Fraiture, C. and Wichelns, D. (2010) Satisfying future demands for agriculture. *Agricultural Water Management* 97, 502–511. doi:10.1016/j.agwat.2009.08.008

de Fraiture, C., Giordano, M. and Yongsong, L. (2008) Biofuels and implications for agricultural water use: blue impacts of green energy. *Water Policy* 10 (Supplement 1), 67–81. doi:10.2166/wp.2008.054

de Wit, M. and Stankiewicz, J. (2006) Changes in surface water supply across Africa with predicted climate change. *Science* 311, 1917–1921. doi:10.1126/science.1119929

Dudgeon, D., Arthington, A.H., Gessner, M.O., Kawabata, Z., Knowler, D.J., Lévêque, C., Naiman, R.J., Prieur-Richard, A., Soto, D., Stiassny, M.L.J. and Sullivan, C.A. (2006) Freshwater biodiversity: importance, threats, status and conservation challenges. *Biological Reviews* 81, 163–182. doi:10.1017/S1464793105006950

EC-FAO Food Security Programme (2008) *An Introduction to the Basic Concepts of Food Security. Food Security Information for Action: Practical Guidelines.* Food Security Information for Action: Practical Guides, European Commission, Brussels and Food and Agriculture Organization of the United Nations, Rome. Available at: http://www.fao.org/docrep/013/al936e/al936e00.pdf (accessed February 2013).

Ericksen, P., Thornton, P., Notenbaert, A., Cramer, L., Jones, P. and Herrero, M. (2011) *Mapping Hotspots of Climate Change and Food Insecurity in the Global Tropics.* CCAFS Report no. 5, CGIAR Research Program on Climate Change, Agriculture and Food Security (CCAFS), Copenhagen, Denmark. Available at: ccafs.cgiar.org/resources/climate_hotspots (accessed December 2012).

FAO (2006) *Global Forest Resources Assessment 2005, Progress Towards Sustainable Forest Management.* FAO Forestry Paper 147, Food and Agriculture Organization of the United Nations, Rome. Available at: www.fao.org/forestry/fra2005 (accessed December 2012).

FAO (2008a) *The State of Food and Agriculture 2008. Biofuels: Prospects, Risks and Opportunities.* Food and Agriculture Organization of the United Nations, Rome. Available at ftp://ftp.fao.org/docrep/fao/011/i0100e/i0100e.pdf (accessed May 2012).

FAO (2008b) *Water for Agriculture and Energy in Africa: Resources and Challenges in the Context of Climate Change. Ministerial Conference [African Union Meeting], Sirte, Libyan Arab Jamahiriya, 15–17 December 2008.* Available at: www.sirtewaterandenergy.org/docs/2009/Sirte_2008_BAK_2.pdf (accessed December 2012).

FAO (2009a) *Climate Change and Bioenergy Challenges for Food and Agriculture. How to Feed the World in 2050: High Level Expert Forum, Rome 12–13 October 2009*. Issues Paper HLEF2050, Food and Agriculture Organization of the United Nations, Rome. Available at: http://www.fao.org/fileadmin/templates/wsfs/docs/Issues_papers/HLEF2050_Climate.pdf (accessed February 2013).

FAO (2009b) *Coping with a Changing Climate: Considerations for Adaptation and Mitigation in Agriculture*. Food and Agriculture Organization of the United Nations, Rome. Available at: www.fao.org/docrep/012/i1315e/i1315e00.htm (accessed December 2012).

FAO (2009c) *Global Agriculture Towards 2050. How to Feed the World in 2050: High Level Expert Forum, Rome 12–13 October 2009*. Issues Paper HLEF2050, Food and Agriculture Organization of the United Nations, Rome. Available at: http://www.fao.org/fileadmin/templates/wsfs/docs/Issues_papers/HLEF2050_Global_Agriculture.pdf (accessed February 2013).

FAO (2010) *The State of Food Insecurity in the World. Addressing Food Insecurity in Protracted Crises*. Food and Agriculture Organization of the United Nations, Rome, Italy.

FAO (2011) *The State of the World's Land and Water Resources for Food and Agriculture (SOLAW) – Managing Systems at Risk (Summary Report)*. Food and Agriculture Organization of the United Nations, Rome and Earthscan, London.

FAO (2012a) FAO Food Price Index. Available at: www.fao.org/worldfoodsituation/wfs-home/foodpricesindex/en/ (accessed 14 April 2012).

FAO (2012b) *The State of World Fisheries and Aquaculture 2012*. FAO Fisheries and Aquaculture Department, Food and Agriculture Organization of the United Nations, Rome. Available at: www.fao.org/docrep/016/i2727e/i2727e00.htm (accessed December 2012).

FAO and PAR (2011) *Biodiversity for Food and Agriculture. Contributing to Food Security and Sustainability in a Changing World. Outcomes of an Expert Workshop Held by FAO and the Platform on [for] Agrobiodiversity Research from 14–16 April 2010 in Rome, Italy*. Food and Agriculture Organization of the United Nations, Rome and Platform for Agrobiodiversity Research, Maccarese, Rome, Italy. Available at: http://agrobiodiversityplatform.org/files/2011/04/PAR-FAO-book_lr.pdf (accessed December 2012).

FAO, WFP and IFAD (2012) *The State of Food Insecurity in the World 2012*. Jointly published by Food and Agriculture Organization of the United Nations, World Food Programme and International Fund for Agricultural Development, Rome. Available at: http://www.fao.org/docrep/016/i3027e/i3027e.pdf (accessed February 2013).

FAOSTATS (2012) FAOSTAT on-line statistical service. Food and Agriculture Organization of the United Nations, Rome. Available at: http://faostat.fao.org/ (accessed February 2013).

Fischlin, A., Midgley, G.F., Price, J.T., Leemans, R., Gopal, B., Turley. C., Rounsevell, M.D.A., Dube, O.P., Tarazona, J. and Velichko, A.A. (2007) Ecosystems, their properties, goods, and services. In: Parry, M.L., Canziani, O.F., Palutikof, J.P., van der Linden, P.J. and Hanson, C.E. (eds) *Climate Change 2007: Impacts, Adaptation and Vulnerability. Contribution of Working Group II to the Fourth Assessment Report of the Intergovernmental Panel on Climate Change, 2007*. Cambridge University Press, Cambridge, UK, pp. 211–272. Available at: www.ipcc.ch/publications_and_data/ar4/wg2/en/contents.html (accessed May 2011).

Foresight (2011) *The Future of Food and Farming: Challenges and Choices for Global Sustainability. Final Project Report*. The Government Office for Science, London.

Giordano, M. and Villholth, K.G. (eds) (2007) *The Agricultural Groundwater Revolution: Opportunities and Threats to Development*. Comprehensive Assessment of Water Management in Agriculture Series 3, CAB International, Wallingford, UK.

Gordon, L.J., Finlayson, C.M. and Falkenmark, M. (2010) Managing water in agriculture for food production and other ecosystem services. *Agricultural Water Management* 94, 512–519. doi:10.1016/j.agwat.2009.03.017

Hannerz, F. (2008*)* Making water information relevant on local to global scale – the role of information systems for integrated water management. PhD thesis, Department of Physical Geography and Quaternary Geology, Faculty of Science, Stockholm University, Stockholm.

Hansen, J., Sato, M. and Ruedy, R. (2012) Perception of climate change. *Proceedings of the National Academy of Sciences of the United States of America* 109, E2415–E2423. doi:10.1073/pnas.1205276109

Harlin, J. (2011) Sustainable water resources management for a green economy: initial findings from the UN Global Survey of Water Resources Management for the Rio+20 Summit. Presentation on 25 August 2011 from the 2011 World Water Week, Stockholm, August 21–27, 2011. Available at: www.

worldwaterweek.org/documents/WWW_PDF/2011/Thursday/K1/Focus-Water-in-a-Green-Economy/ Sustainable-water-resources-management-for-a-green-economy.pdf (accessed December 2012).

Herrero, M., Thornton, P.K., Gerber, P. and Reid, R.S. (2009) Livestock, livelihoods and the environment: understanding the trade-offs. *Current Opinion in Environmental Sustainability* 1, 111–120. doi:10.1016/j. cosust.2009.10.003

Herrero, M. *et al.* (2010) Smart investments in sustainable food production: revisiting mixed crop–livestock systems. *Science* 327, 822–825. doi:10.1126/science.1183725

Hoanh, C.T., Szuster, B.W., Kam, S.P., Ismail, A.M. and Noble, A.D. (eds) (2010) *Tropical Deltas and Coastal Zones: Food Production, Communities and Environment at the Land–Water Interface.* Comprehensive Assessment of Water Management in Agriculture Series 9. CAB International, Wallingford, UK, in association with International Water Management Institute (IWMI), Colombo, WorldFish, Penang, Malaysia, International Rice Research Institute (IRRI), Los Banos, Philippines, FAO Regional Office for Asia and the Pacific, Bangkok, and CGIAR Challenge Programme on Water and Food (CPWF), Colombo.

Hoekstra, A.Y. and Mekonnen, M.M. (2012) The water footprint of humanity. *Proceedings of the National Academy of Sciences of the United States of America* 109, 3232–3237. doi:10.1073/pnas.1109936109

Huchet-Bourdon, M. (2011) *Agricultural Commodity Price Volatility: An Overview.* OECD Food, Agriculture and Fisheries Working Papers, No. 52, Organisation for Economic Co-operation and Development, Paris. doi:10.1787/5kg0t00nrthc-en

IFAD (2010) *Rural Poverty Report 2011. New Realities, New Challenges: New Opportunities for Tomorrow's Generation.* International Fund for Agricultural Development, Rome.

IPCC (2012) *Managing the Risks of Extreme Events and Disasters to Advance Climate Change Adaptation. A Special Report of Working Groups I and II of the Intergovernmental Panel on Climate Change.* Edited by Field, C.B., Barros, V., Stocker, T.F., Dahe, Q., Dokken, D.J., Ebi, K.L., Mastrandrea, M.D., Mach, K.J., Plattner, G.-K., Allen, S.K., Tignor, M. and Midgley, P.M.. Cambridge University Press, Cambridge, UK.

Janzen, H.H. (2004) Carbon cycling in earth systems – a soil science perspective. *Agriculture, Ecosystems and Environment* 104, 399–417. doi:10.1016/j.agee.2004.01.040

Kearney, J. (2010) Food consumption trends and drivers. *Philosophical Transactions of the Royal Society B* 365, 2793–2807. doi:10.1098/rstb.2010.0149

Keys, P., Barron, J. and Lannerstad, M. (2012) *Releasing the Pressure: Water Resource Efficiencies and Gains for Ecosystem Services.* United Nations Environment Programme (UNEP), Nairobi and Stockholm Environment Institute, Stockholm.

Knox, J.W., Hess, T.M., Daccache, A. and Perez Ortola, M. (2011) *What Are the Projected Impacts of Climate Change on Food Crop Productivity in Africa and South Asia? DFID Systematic Review, Final Report.* Cranfield University, Cranfield, UK, on commission from DFID (Department for International Development, London).

Le Quesne, T. *et al.* (2010) *Flowing Forward: Freshwater Ecosystem Adaptation to Climate Change in Water Resources Management and Biodiversity Conservation.* Water Working Notes No. 28, The World Bank, Washington, DC.

Lenton, M.T., Held, H., Kriegler, E., Hall, J.W., Lucht, W., Rahmstorf, S. and Schellnhuber, H.J. (2008) Tipping elements in the earth's climate system. *Proceedings of the National Academy of Sciences of the United States of America* 105, 1786–1793. doi:10.1073/pnas.0705414105

Liapis, P. (2012) *Structural Change in Commodity Markets: Have Agricultural Markets Become Thinner?* OECD Food, Agriculture and Fisheries Working Papers, No. 54, Organisation for Economic Co-operation and Development, Paris. doi:10.1787/5k9fp3zdc1d0-en

Lobell, D.B., Burke, M.M., Tebaldi, C., Mastrandrea, M.D., Falcon, W.P. and Naylor, R.L. (2008) Prioritizing climate change adaptation needs for food security in 2030. *Science* 319, 607–610. doi:10.1126/ science.1152339

Lowder, S.K. and Carisma, P. (2011) *Financial Resource Flows to Agriculture: A Review of Data on Government Expenditures, Official Development Assistance and Foreign Direct Investment.* ESA Working Paper 11-19, Agricultural Development Economics Division, Food and Agriculture Organization of the United Nations, Rome. Available at: www.fao.org/docrep/015/an108e/an108e00.pdf (accessed December 2012).

Lundqvist, J., Barron, J., Berndes, G., Berntell, A., Falkenmark, M., Karlberg, L. and Rockström, J. (2007) Water pressure and increases in food and bioenergy demand – implications of economic growth and options for de-coupling. In: Falkenmark, M., Karlberg, L. and Rockström, J. (eds) *Scenarios on Economic Growth and Resource Demand. Background Report to the Swedish Environmental Advisory Council,*

Memorandum 2007:1. Swedish Environmental Advisory Council, Stockholm, pp. 55–151. Available at: http://www.sou.gov.se/mvb/pdf/WEBB-%20PDF.pdf (accessed February 2013).

Lutz, W. and Samir, K.C. (2010) Dimensions of global population projections: what do we know about future population trends and structures? *Philosophical Transactions of the Royal Society B* 365, 2779–2791. doi:10.1098/rstb.2010.0133

Makarieva, A.M., Gorshkov, V.G., Sheil, D., Nobre, A.D. and Li, B.-L. (2010) Where do winds come from? A new theory on how water vapor condensation influences atmospheric pressure and dynamics. *Atmospheric Chemistry and Physics Discussions* 10, 24015–24052. doi:10.5194/acpd-10-24015-2010

McCartney, M. and Smakhtin, V. (2010) *Water Storage in an Era of Climate Change: Addressing the Challenge of Increasing Rainfall Variability*. IWMI Blue Paper, International Water Management Institute, Colombo.

McIntyre, B.D., Herren, H.R., Wakhungu, J. and Watson, R.T. (eds) (2008) *Agriculture at a Crossroads. International Assessment of Agricultural Knowledge, Science and Technology for Development (IAASTD): Global Report*. Island Press, Washington, DC. Available at: http://www.agassessment.org/reports/IAASTD/EN/Agriculture%20at%20a%20Crossroads_Global%20Report%20(English).pdf (accessed February 2013).

Metz, B., Davidson, O.R., Bosch, P.R., Dave, R. and Meyer, L.A. (eds) (2007) *Climate Change 2007: Mitigation of Climate Change. Contribution of Working Group III to the Fourth Assessment Report of the Intergovernmental Panel on Climate Change, 2007*. Cambridge University Press, Cambridge, UK. Available at: www.ipcc.ch/publications_and_data/ar4/wg3/en/contents.html (accessed December 2012).

Millennium Ecosystem Assessment (2005a) *Ecosystems and Human Well-being: Synthesis. A Report of the Millennium Ecosystem Assessment*. World Resources Institute and Island Press, Washington, DC. Available at: www.maweb.org/documents/document.356.aspx.pdf (accessed February 2013).

Millennium Ecosystem Assessment (2005b) *Ecosystems and Human Well-being: Wetlands and Water – Synthesis. A Report of the Millennium Ecosystem Assessment*. World Resources Institute, Washington, DC. Available at: www.maweb.org/documents/document.358.aspx.pdf (accessed February 2013).

Molden, D. (ed.) (2007) *Water for Food, Water for Life: Comprehensive Assessment of Water Management in Agriculture*. Earthscan, London, in association with International Water Management Institute (IWMI), Colombo.

Mosier, A.R., Duxbury, J.M., Freney, J.R,. Heinemeyer, O., Minami, K. and Johnson, D.E. (1998) Mitigating agricultural emissions of methane. *Climatic Change* 40, 39–80. doi:10.1023/A:1005338731269

Nellemann, C., MacDevette, M., Manders, T., Eickhout, B., Svihus, B., Prins, A.G. and Kaltenborn, B.P. (eds) (2009) *The Environmental Food Crisis – The Environment's Role in Averting Future Food Crises. A UNEP Rapid Response Assessment*. United Nations Environment Programme, GRID-Arendal, Norway. Available at: http://www.grida.no/files/publications/FoodCrisis_lores.pdf (accessed February 2013).

Nilsson, C., Reidy, C.A., Dynesius, M. and Revenga, C. (2005) Fragmentation and flow regulation of the world's large river systems. *Science* 308, 405–408.

O'Mara, F.P. (2011) The significance of livestock as a contributor to global greenhouse gas emissions today and in the near future. *Animal Feed Science and Technology* 166, 7–15.

OECD and FAO (2011) *OECD-FAO Agricultural Outlook 2011–2020*. Organisation for Economic Co-operation and Development, Paris and Food and Agriculture Organization of the United Nations, Rome. Available at: www.fao.org/fileadmin/user_upload/newsroom/docs/Outlookflyer.pdf (accessed December 2012).

OECD and FAO (2012) *OECD-FAO Agricultural Outlook 2012–2021*. Organisation for Economic Co-operation and Development, Paris and Food and Agriculture Organization of the United Nations, Rome. doi:10.1787/agr_outlook-2012-en

OECD-DAC (2010) *Measuring Aid to Agriculture*. OECD Development Assistance Committee (DAC), Organisation for Economic Co-operation and Development, Paris. Available at: http://www.oecd.org/investment/stats/44116307.pdf (accessed February 2013).

OECD-DAC (2011) *Aid to Agriculture and Rural Development*. OECD Development Assistance Committee (DAC), Organisation for Economic Co-operation and Development, Paris. Available at: http://www.oecd.org/development/stats/49154108.pdf (accessed February 2013).

Oenema, O., Wrage, N., Velthof, G.L., van Groenigen, J.W., Dolfing, J. and Kuikman, P.J. (2005) Trends in global nitrous oxide emissions from animal production systems. *Nutrient Cycling in Agroecosystems* 72, 51–65. doi:10.1007/s10705-004-7354-2

Ong, C.K., Black, C.B. and Muthuri, C.W. (2006) Modifying Forestry and Agroforestry to Increase Water Productivity in the Semi-arid Tropics. CAB Reviews: Perspectives in Agriculture, Veterinary Science, Nutrition and Natural Resources 1, no. 065. CAB International, Wallingford, UK. Available at: www.cabi.

org/cabreviews/default.aspx?site=167&page=4051&LoadModule=Review&ReviewID=26188 (accessed May 2012).

Ostrom, E. (2009) A general framework for analyzing sustainability of social-ecological systems. *Science* 325, 419–422. doi:10.1126/science. 1172133

Oxfam (2011) *Growing a Better Future: Food Justice in a Resource-constrained World.* Oxfam International and Oxfam GB, Oxford, UK. Available at http://www.oxfam.org/grow/reports/growing-better-future (accessed December 2012).

Parry, M.L., Canziani, O.F., Palutikof, J.P., van der Linden, P.J. and Hanson, C.E. (eds) (2007) *Climate change 2007: impacts, adaptation and vulnerability. Contribution of Working Group II to the Fourth Assessment Report of the Intergovernmental Panel on Climate Change, 2007.* Cambridge: Cambridge University Press Cambridge, UK. Available at: www.ipcc.ch/publications_and_data/ar4/wg2/en/contents.html (accessed December 2012).

Peden, D., Tadesse, G. and Misra, A.K. *et al.* (2007) Water and livestock for human development. In: Molden, D. (ed.) *Water for Food, Water for Life: Comprehensive Assessment of Water Management in Agriculture.* Earthscan, London, in association with International Water Management Institute (IWMI), Colombo, pp. 485–514.

Pelletier, N. and Tyedmers, P. (2010) Forecasting potential global environmental costs of livestock production 2000–2050. *Proceedings of the National Academy of Sciences of the United States of America* 107, 18371–18374.

Rijsberman, F.R. (2006) Water scarcity: fact or fiction? *Agricultural Water Management* 80, 5–22. doi:10.1016/j.agwat.2005.07.001

Rockström, J., Falkenmark, M., Karlberg, L., Hoff, H., Rost, S. and Gerten, D. (2009a) Future water availability for global food production: the potential of green water to build resilience to global change. *Water Resources Research* 44(7): W00A12. doi.org/10.1029/2007WR006767

Rockström, J. *et al.* (2009b) A safe operating space for humanity. *Nature* 461, 472–475. doi:10.1038/461472a

Schuman, G.E., Janzen, H.H. and Herrick, J.E. (2002) Soil carbon dynamics and potential carbon sequestration by rangelands. *Environmental Pollution* 116, 391–396. doi:10.1016/S0269-7491(01)00215-9

Sharma, H.P. (2012) Migration, remittance and food security: a complex relationship. *The Development Review – Beyond Research* 11, 42–67.

Smith, K.A. and Conen, F. (2004) Impacts of land management on fluxes of trace greenhouse gases. *Soil Use and Management* 20, 255–263. doi:10.1111/j.1475-2743.2004.tb00366.x

Steffen, W. *et al.* (2011) The Anthropocene: from global change to planetary stewardship. *Ambio* 40, 739–761. doi:10.1007/s13280-011-0185-x

Steinfeld, H., Gerber, P., Wassenaar, T., Castel, V., Rosales, M. and de Haan, C. (2006) *Livestock's Long Shadow: Environmental Issues and Options.* Food and Agriculture Organization of the United Nations, Rome. Available at: www.fao.org/docrep/010/a0701e/a0701e00.HTM (accessed December 2012).

Tacoli, C. (2011) *Not Only Climate Change: Mobility, Vulnerability and Socio-economic Transformations in Environmentally Fragile Areas of Bolivia, Senegal and Tanzania.* Human Settlements Working Paper Series 28, International Institute for Environment and Development, London. Available at: http://pubs.iied.org/10590IIED.html (accessed December 2012).

Tilman, D., Fargione, J., Wolff, B., D'Antonio, C., Dobson, A., Howarth, R., Schindler, S., Schlesinger, W.H., Simberloff, D. and Swackhamer, D. (2001) Forecasting agriculturally driven global environmental change. *Science* 292, 281–284.

Tilman, D., Balzer, C., Hill, J. and Befort, B.L. (2011) Global food demand and the sustainable intensification of agriculture. *Proceedings of the National Academy of Sciences of the United States of America* 108, 20260–20264. doi:1073/pnas.1116437108

Turral, H., Svendsen, M. and Faures, J.M. (2010) Investing in irrigation: reviewing the past and looking to the future. *Agricultural Water Management* 97, 551–556. doi:10.1016/j.agwat.2009.07.012

UN (2011) *The Millennium Development Goals Report 2011.* United Nations, New York. Available at: http://mdgs.un.org/unsd/mdg/Resources/Static/Products/Progress2011/11-31339%20(E)%20MDG%20Report%202011_Book%20LR.pdf (accessed February 2013).

UNEP (2007) *Global Environment Outlook. GEO-4, Environment for Development.* United Nations Environment Programme, Nairobi. Available at: www.unep.org/geo/geo4.asp (accessed February 2010).

UNEP (2009) *Ecosystem Management Programme. A New Approach to Sustainability.* United Nations Environment Programme, Nairobi. Available at: http://www.unep.org/ecosystemmanagement/Portals/7/Documents/EMP-Booklet.pdf (accessed February 2013).

UNEP (2010) *Blue Harvest: Inland Fisheries as an Ecosystem Service.* WorldFish, Penang, Malaysia and
    United Nations Environment Programme, Nairobi.

von Braun, J. (2007) *The World Food Situation. New Driving Forces and Required Actions.* Food Policy
    Report, International Food Policy Research Institute (IFPRI), Washington, DC. Available at: www.ifpri.
    org/sites/default/files/pubs/pubs/fpr/pr18.pdf (accessed December 2012).

Vörösmarty, C.J., McIntyre, P.B., Gessner M.O., Dudgeon, D., Prusevich, A., Green, P., Glidden, S., Bunn,
    S.E., Sullivan, C.A., Reidy Liermann, C. and Davies, P.M. (2010) Global threats to human water security
    and river biodiversity. *Nature* 467, 555–561. doi:10.1038/nature09440

Washington, R., New, M., Hawcroft, M., Pearce, H., Rahiz, M. and Karmacharya, J. (2012*) Climate Change
    in CCAFS Regions: Recent Trends, Current Projections, Crop–Climate Suitability, and Prospects for
    Improved Climate Model Information.* CGIAR Research Program on Climate Change, Agriculture and
    Food Security (CCAFS), Copenhagen.

Woomer, P.L., Touré, A. and Sall, M. (2004) Carbon stocks in Senegal's Sahel transition zone. *Journal of Arid
    Environments* 59, 499–510. doi:10.1016/j.jaridenv.2004.03.027

World Bank (2012) World development indicators. Available at http://data.worldbank.org/indicator/ (accessed
    October 2012).

WRI (2005) *World Resources 2005: The Wealth of the Poor – Managing Ecosystems to Fight Poverty.* World
    Resources Institute, Washington, DC,. Available at: http://www.wri.org/publication/world-resources-
    2005-wealth-poor-managing-ecosystems-fight-poverty (accessed February 2013).

WWAP (2009) *The United Nations World Water Development Report 3 (WWDR3). Water in a Changing
    World.* World Water Assessment Programme, United Nations Educational, Scientific and Cultural
    Organization (UNESCO), Paris and Earthscan, London. Available at: http://unesdoc.unesco.org/
    images/0018/001819/181993e.pdf#page=5 (accessed February 2013).

WWAP (2012) *The United Nations World Water Development Report 4 (WWDR4): Managing Water Under
    Uncertainty and Risk (Vol. 1), Knowledge Base (Vol. 2)* and *Facing the Challenges (Vol. 3).* World Water
    Assessment Programme, United Nations Educational, Scientific and Cultural Organization (UNESCO),
    Paris.

3 Water-related Ecosystem Services and Food Security

David Coates,[1]* Petina L. Pert,[2] Jennie Barron,[3] Catherine Muthuri,[4] Sophie Nguyen-Khoa,[5] Eline Boelee[6] and Devra I. Jarvis[7]

[1] *Secretariat of the Convention on Biological Diversity (CBD), Montreal, Canada;* [2] *Commonwealth Scientific and Industrial Research Organisation CSIRO, Ecosystem Sciences, Cairns, Queensland, Australia;* [3] *Stockholm Environment Institute, University of York, UK and Stockholm Resilience Centre, Stockholm University, Stockholm, Sweden;* [4] *World Agroforestry Centre (ICRAF), Nairobi, Kenya;* [5] *World Water Council (WWC), Marseille, France;* [6] *Water Health, Hollandsche Rading, the Netherlands;* [7] *Bioversity International, Rome, Italy*

Abstract

The ecosystem setting of both agriculture and water provides a conceptual framework for managing the needs of agriculture for water and the impacts of water upon agriculture. Water underpins all benefits (ecosystem services) that ecosystems provide, including all agricultural production. The availability of water, in terms of both its quantity and quality, is also influenced heavily by ecosystem functioning. Understanding this relationship of water, ecosystems and their services with agriculture is at the heart of understanding, and therefore managing, water and food security. There are opportunities to move beyond seeing the agriculture–ecosystem–water interface as one of conflict and trade-offs, towards simultaneously achieving both increases in sustainable food production and improvements in the delivery of other ecosystem benefits by agriculture through more widespread adoption of ecosystem-based solutions. These concepts and approaches are explained briefly here as an introduction to understanding the interlinkages between ecosystem services, water and food security in subsequent chapters of the book.

Background

The water cycle is a biophysical process, heavily influenced by ecosystem functioning. The healthy functioning of ecosystems underpins a multitude of benefits (services) derived from ecosystems. Water is a critical component in maintaining these functions, while keeping them resilient to change (Costanza *et al.*, 1997). The presence and absence of water in the landscape very often determines the characteristics of several supporting and regulating functions, e.g. preserving nutrients and removing pollutants (Falkenmark, 2003).

* E-mail: david.coates@cbd.int

© CAB International 2013. *Managing Water and Agroecosystems for Food Security* (ed. E. Boelee)

This chapter provides an introduction to how agriculture depends upon, and influences, water in this ecosystem context. Importantly, this context brings with it opportunities for managing ecosystems as solutions to achieve water and food security, which are further developed in subsequent chapters in this volume, notably in Chapters 4 and 9.

The water cycle at the agroecosystem scale is illustrated in Fig. 3.1. Water is a key factor to be managed to enhance agricultural benefits, whether in rainfed or in irrigated farming systems. In rainfed farming systems, management aims to maximize soil infiltration of rainwater and soil water holding capacity or, in some cases, to drain excess water to ensure good growth. In irrigation, the same management aim is met from water derived from external sources (surface or groundwater sources) at timely intervals for the crop.

The implications of considering water in this ecosystem context are twofold. First, as explained further below, water underpins many ecosystem benefits, food production being only one. Although it has long been established that using water in agriculture has implications for other uses, there remains, in many circles, limited understanding of how these impacts are delivered, their importance and how they can be managed. Secondly, water management policies in agriculture can be dominated by considering visible surface water and groundwater (e.g. irrigation), whereas the less visible parts of the water cycle (e.g. land cover and cycling through soils) are important and can often be underemphasized. Molden (2007), for example, noted that while potential productivity gains are available in irrigated agriculture, perhaps the biggest opportunities lie with rainfed agriculture, which largely involves improving rainwater retention by soils (see Chapter 8). Some ecosystem-driven aspects of the water cycle that merit better attention include:

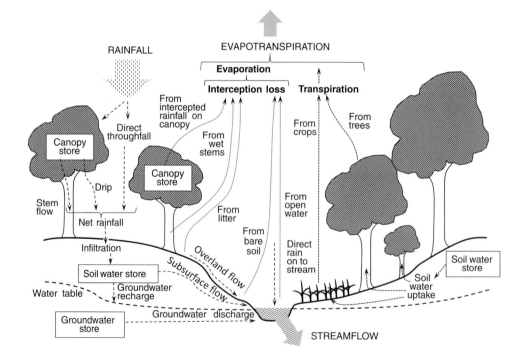

Fig. 3.1. The hydrological cycle in an agroecosystem.

- The role of wetlands in regulating surface, and in some cases groundwater, flows (see Chapter 7).
- Soil functionality, particularly in retaining water not only as water security for crops but also as a major component of the overall water cycle (desertification, for example, is a process essentially driven by loss of water from soil; see also Chapter 4).
- The importance of vegetation (land cover) as a major component of the water cycle (Box 3.1).
- How ecosystems can be regarded as a 'natural water infrastructure', which functions in a similar fashion to human-built (physical) infrastructure and therefore offers options for addressing water management needs.

Water management in agriculture thus essentially requires a very comprehensive approach. In many situations, focusing too much on managing visible surface water results in water 'supply' (in terms of the absolute quantity of physical surface water) being considered an unmanageable variable (driven essentially by unpredictable rainfall). In fact, this is far from the case, as Box 3.1 illustrates.

The ecosystem context of water presents a paradigm shift in how we think about the water–food–environment interface. Historically the water–environment interface has been largely one of conflict in which the 'environment' (or ecosystem!) has been regarded as an unfortunate but necessary victim of development. An alternative approach is to view water management as the management of water use

and ecosystems in order to deliver multiple ecosystem benefits in a mutually supporting way (Fig. 3.2).

Agroecosystems

Agriculture is an ecosystem management activity from which primary and secondary agricultural products are appropriated by humans (Fresco, 2005). An 'ecosystem' can be defined as a dynamic complex of plants, animals, microorganisms and their non-living environment, of which people are an integral part (UNEP, 2009). All agricultural activities depend on a functioning ecosystem, for example healthy soil or the presence of pollinators, but can also have impacts on the ecosystem beyond the immediate interests of agriculture, for example downstream water pollution. Defining the management components of ecosystems is largely a matter of scale. Discrete ecosystem types can often be identified (for example, soils, wetlands, mountains, drylands, forests) but, although some management activities might focus on these discrete elements (for example, managing soil in a field), the reality is that all these components are interconnected, and particularly so through water (see Figs 3.1 and 3.3).

In this book, we refer to areas where agriculture is the dominant land use activity as 'agroecosystems' in order to recognize both the dependency of agriculture on the ecosystem and its setting within the broader landscape (Conway, 1987). Certain components of agroecosystems are particularly relevant to the

---

Box 3.1. The importance of vegetation in managing water

Deforestation can decrease regional rainfall through the loss of cloud-forming evapotranspiration from the forest. Local climate then becomes drier, thereby accelerating ecosystem change. Science suggests that in the Amazon, for example, feedback loops mean that apparently moderate deforestation of 20% could mean that a tipping point is reached beyond which forest ecosystems collapse across the entire basin (Vergara and Scholz, 2011). This would have devastating impacts on water security and other ecosystem services that would reach far beyond the Amazon Basin itself, including through impacts on regional agriculture and global carbon storage. Worryingly, deforestation in the Amazon is already of the order of 18% (Vergara and Scholz, 2011).

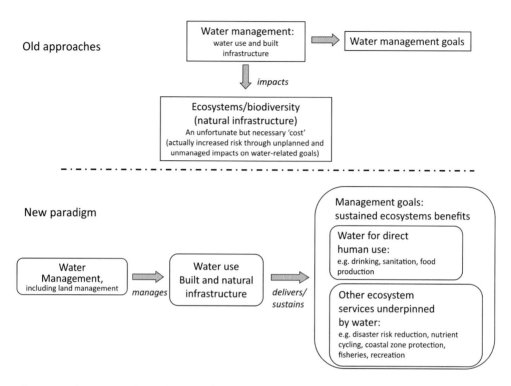

Fig. 3.2. Evolving approaches to the water–ecosystem interface.

scope of water and food security. These include: open water bodies (such as wetlands, rivers and lakes) that can supply water to agriculture but also compete with agriculture over water, and are affected by agrochemicals such as fertilizers and pesticides; and soils, which are the immediate source of water for most crops. Most agroecosystems, certainly at the larger scale, contain a mosaic of multiple land use types. These can vary from, for example, large expanses of natural or managed forest and plantations, such as coffee and rubber plantations, through to hedgerows used to divide fields or protect riverbanks, interspersed with human settlements and transport infrastructure. The combinations of land use types and activities, together with the topographic and climate setting, results in certain clearly identifiable agroecosystem types, such as the rice systems in South-east Asia or the vast cereal plains of the Midwest USA. Each of these has its own particular issues of vulnerability and management.

In order to understand ecosystem services, this book considers a continuum of ecosystem conditions from undisturbed pristine ('nature') areas to highly managed and altered systems. While the condition of an ecosystem can greatly determine its ability to function, and therefore provide services (benefits), a highly modified area (e.g. intensively monocropped farmland) is still an ecosystem. Debate about 'natural' versus 'managed' ecosystems is largely redundant as approaching 90% of the earth's terrestrial surface is influenced, in at least some respect, by human activity (Ellis and Ramankutty, 2008; Ellis, 2011). Almost all so-called natural ecosystems are influenced by people as hunters, gatherers and foragers actively managing the landscape to facilitate their harvesting of food and other useful products (Bharucha and Pretty, 2010). If climate change influences are included, there is arguably no 'natural' area left at all. The focus needs to be: what services do we want the landscape to provide (including, where desired,

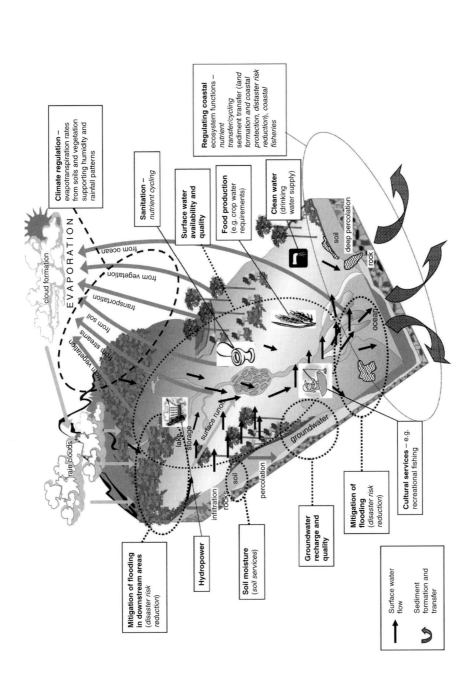

Fig. 3.3. A conceptual framework illustrating the water cycle and ecosystem services in a simplified landscape setting. The figure lists in italics (in the boxes) some of the water-related benefits to people (ecosystem services) that ecosystem functioning provides. In reality, the various services illustrated, and others, are more dispersed, interconnected and affected by land- and water-use activities (not shown in full) (based on MRC, 2003).

the various benefits of 'nature') and how can it be managed in order to sustain the desired supply of those services.

An agroecosystem perspective also helps to give value to ecosystem services (see next section). According to FAO, agroecosystems are ecosystems in which humans have exercised a deliberate selectivity and modified the composition of existing fauna and flora for agricultural purposes (OECD, 2011). Together, these agroecosystems cover over a third of the total terrestrial area. Agroecosystems both provide and rely upon important ecosystem services (Zhang et al., 2007). For example, sustainable flooded rice–aquaculture systems build upon the disease and pest regulation and nutrient cycling services provided by a healthy aquatic ecosystem to underpin food production (of both rice and fish), and also provide nutrient cycling and water regulating services beyond agriculture.

Agriculture in the ecosystem context has been explored in detail by the International Assessment of Agricultural Knowledge, Science and Technology for Development (IAASTD) (McIntyre et al., 2008). In addition to providing food (for a nutritious diet), fibres, fuel, fodder and related employment (agriculture is a major employer: globally, 40% of livelihoods depend on it; McIntyre et al., 2008), agriculture also delivers a variety of other goods and ecosystem services, and fulfils various social and cultural roles. For example, farmers are key stakeholders in managing landscapes and the cultural benefits associated with them.

Aquaculture has been an integral part of many agroecosystems for thousands of years, producing additional food and cash to supplement crop and livestock production, making more efficient use of feed and fertilizer inputs, and facilitating nutrient retention and recycling from manure, agricultural and food processing by-products, and domestic waste-water. It is especially crucial for poor women who often have few other income-earning opportunities (UNFPA, 2009).

The characteristics of various types of agroecosystems are determined by environ-mental factors (e.g. climate, topography, water availability and soil type in which they are situated) but also by the socio-economic setting, including demand for products, traditional and

historical practices, supporting policies, technical capacity and financial capital availability. These, together with other factors, determine the farming systems in place and the way they evolve. Water availability plays a key role in farmers' risk management strategies. Water availability varies naturally throughout the year, and between years, in most farming systems. The inability to predict the exact amount available throughout a growing season results in significant uncertainties for farmers. However, the highest level of risk is associated not with mean supply but with extremes in water availability from both flooding and drought. Farmers adopt various strategies to cope with such risks; for example, through crop diversification, so they have at least some production in the event of an extreme event. None the less, the very existence of water-related risks is a major constraint to investment aimed at increasing agricultural productivity, particularly for poor, and therefore more vulnerable, farmers.

Ecosystem Services

The benefits that we as humans derive from ecosystems, such as timber, food, water and climate regulation, are referred to collectively as ecosystem services (Millennium Ecosystem Assessment, 2005a); further details are provided in Chapters 4 and 9. The concept of ecosystem services is used to analyse trade-off scenarios when human well-being and ecological sustainability need to be addressed simultaneously. The ecosystem perspective aims to bridge interdisciplinary gaps between fields as far apart as religion and biology, political science and geology or engineering and biodiversity, thereby addressing the system comprehensively.

The availability of water at any time or place, in terms of both its quantity and quality, is also a service provided by ecosystems, and one of obvious importance to agriculture. Because water is required for ecosystems to function, all ecosystem services (excepting some of those provided by marine environ-ments, particularly oceans) are underpinned by fresh water (Aylward et al., 2005; UCC-Water, 2008).

Ecosystem services can be grouped into four different types (Millennium Ecosystem Assessment, 2005a), as follows:

- *Provisioning services* are essentially the tangible products (or goods) that are used directly by humans. These are among the most recognizable in terms of human use and are thus most frequently monetized but are not necessarily the most valuable. Relevant examples include freshwater (directly used, e.g. for drinking), energy from hydropower and all food (including all the products of agriculture, livestock rearing, forestry, fisheries and wild-caught products such as bushmeat). Globally, provisioning services have been maximized, particularly by agriculture, at the expense of reductions in other services (listed below), resulting in a serious imbalance (Millennium Ecosystem Assessment, 2005a).
- *Regulatory (or 'regulating') services* are the benefits that ecosystems provide in terms of regulating ecosystem-dependent processes. Relevant examples include: climate regulation (including precipitation), water regulation (i.e. hydrological flow), water purification and waste treatment, erosion regulation and water-related natural hazard regulation. Such services are sometimes less tangible at farm and field scales, and can be more difficult to assess economically (although there are exceptions; natural hazard regulation, for example, is more easily assessed because the impacts of disasters can often be quantified in fairly standard economic terms). In some instances, these services can be replaced by technology but often at a higher cost than that of maintaining the original service (Cairns, 1995): e.g. any infrastructure or operational costs in treating water to make it potable are essentially expenditures on replacing the original water purification and supply functions of ecosystems, which previously provided this service free.
- *Cultural services* include the spiritual and inspirational, religious, recreational, aesthetic and educational benefits that people derive directly or indirectly from ecosystems: for example, the recreational benefit of a lake for fishing. Some are more easy to value (e.g. through amounts spent on recreation, including transport and accommodation costs), but others are less tangible and often difficult to quantify or monetize. Nevertheless, the importance of cultural service values should not be underestimated; they represent some of the clearest examples of the pitfalls of monetized economic valuations. An example is the case of pastoral livestock, where cultural values can override economic values in terms of development and land management, and include 'antiquity, role in the agricultural systems, farming techniques, role in landscape, gastronomy, folklore and handicrafts' (Gandini and Villa, 2003).
- *Supporting services* are those that underpin broader ecosystem functioning and hence contribute to sustaining other services. Examples include soil formation and nutrient cycling, both of which are essentially water based and aquatic ecosystem driven processes.

The tendency to maximize provisioning services at the expense of the other services is partly because most provisioning services are marketed and the market value does not reflect the external costs of impacts on other services (Millennium Ecosystem Assessment, 2005a). This can particularly affect poor people as they are often more closely and directly dependent on several ecosystem services, and are affected the most severely when services degrade, for example, the availability of clean drinking water or of firewood (WRI *et al.*, 2008). This situation is likely to get worse under the influence of population growth, continued abuse of ecosystem services and global climate change (Mayers *et al.*, 2009). Water-related ecosystem services, derived essentially from how ecosystems underpin the water cycle, are important renewable resources. They provide many promising solutions to the need to achieve sustainable agriculture: for example, restoring soil ecosystem services can be key to sustaining water availability for crops, reinstating nutrient retention in soil, and cycling and reducing erosion and rainfall runoff (hence reducing water-related impacts downstream); examples of such approaches are provided in subsequent chapters. However, enacting these solutions requires good governance, and more research is needed on how to secure the regulatory and supporting services of ecosystems in order to help with poverty reduction (Mayers *et al.*, 2009).

Changes in the local or regional availability of water, and its quality, whether due to agricultural or any other influence, consequently have implications for the delivery of ecosystem services at local and regional scales (Fig. 3.3). The management of the interdependency between water and ecosystem services, underpinned by ecosystem functions, and illustrated in Fig. 3.3, is at the heart of meeting two of the major challenges facing agriculture – water security for food security and water security for other purposes – and is therefore a core subject of this book.

Although many ecosystem services are known to be important to agriculture, the mechanistic details of their provision, or reduction, remain poorly understood (Kremen, 2005), and we lack ways to quantify many ecological services in a manner similar to measures of marketed goods and services in the economy (Dale and Polasky, 2007). Moreover, the provisioning services that we can measure depend upon a wide variety of supporting and regulatory services, such as soil fertility and pollination (Millennium Ecosystem Assessment, 2005a) that determine the underlying biophysical capacity of agro-ecosystems (Wood *et al.*, 2000). Agro-ecosystems can also be affected by activities beyond agriculture, such as impacts on water from non-agricultural sources, which might reduce agricultural productivity or increase production costs (Zhang *et al.*, 2007).

Ecosystem services are central to the well-being of all humans but are particularly directly relevant to the livelihoods of the rural poor. For example, while agriculture, forests and other ecosystems together comprise 6% of the gross domestic product (GDP) in Brazil and 11% of that in Indonesia, these ecosystem services contribute more than 89% of the GDP to poor households in Brazil and 75% to those in Indonesia, thus benefiting 18 and 25 million people in Brazil and Indonesia, respectively (TEEB, 2010). Hence, there is significant potential to contribute to poverty reduction through the better management of agro-ecosystems.

Balancing Multiple Ecosystem Services

One of the main challenges to achieving water and food security is land and water management that balances the continued delivery of the full suite of necessary ecosystem services required to sustain overall well-being. Because these ecosystem services are largely interdependent, and in particular because of the interlinkages that occur through water use and impacts (Fig. 3.3), there is often, but not always, a trade-off element in decision making. Trade-offs, though, are not necessarily linear (an increase in one service does not necessarily decrease another by an equal amount), and there is room to move the ecosystem services debate on: from a 'trade-off mentality' to one of achieving efficient use of ecosystems. For example, through identifying approaches that achieve food security objectives and at the same time meet other sustainable development objectives for water.

Simplistically, there are two aspects of managing ecosystem services at the water–food interface:

- First, managing those water-related eco-system services that are required in order to sustain increased agricultural productivity (e.g. improved water retention by soils). With these, there is an incentive for agricultural policies, and in particular for farmers, to manage these services.
- Secondly, managing those services that are under the influence of agriculture but do not benefit agricultural communities directly ('downstream impacts'). Here, there are limited or negative incentives for agriculture, and especially for farmers, to manage such impacts. For example, asking farmers to manage land better (to benefit downstream users, perhaps through improved water quality) is unlikely to be popular with them if they incur increased production costs. Solutions to this dilemma, other than regulation, include: (i) identifying behavioural change that benefits both farmers and other stake-holders (win–win outcomes); and (ii) in par-ticular, identifying ways and means to improve incentives for farmers to change

their behaviour through payments for ecosystem services (as discussed further in Chapter 9).

Improved knowledge of the whole range of ecosystem services, their benefits (values) and costs (social, financial, water) can help to achieve better decisions on water and land use (Millennium Ecosystem Assessment, 2005a; TEEB, 2010). Well-balanced decisions, including trade-offs where necessary, can often enhance overall ecosystem services without sacrificing productivity (Millennium Ecosystem Assessment, 2005a; Bennett *et al.*, 2009). The separation of ecosystem services into market and non-market goods leads to a disconnect between economics and environmental sustainability because variations in non-market goods are not reflected in economic pricing and monetary flows (Wilson and Carpenter, 1999; Millennium Ecosystem Assessment, 2005a); there are no direct market-based economic incentives to sustain important ecosystem services if these are not valued, priced and traded. An example is that few, if any, stakeholders pay the full environment costs of water use. Groundwater recharge and climate regulation are other examples where an individual's benefits from these services are not directly linked to the cost of using them (Millennium Ecosystem Assessment, 2005b).

By estimating the value of an ecosystem's market and non-market goods, hidden social and environmental costs and benefits can be made visible (Wilson and Carpenter, 1999). Some regulating and supporting services can be brought into markets and evaluated in financially driven decision-making processes by exploring the costs of substituting for them. For example, a watershed's purification functions can be monetized by comparison with the cost of substituting a water treatment facility to fulfil these needs for a community. Some ecologists, however, have argued against this logic, suggesting that humans cannot fully substitute for the functions of these regulating systems, especially as they contribute to multiple services and biodiversity (Ehrlich and Mooney, 1983). This dilemma is one of the central issues of debate on the valuation of ecosystem services (Ehrlich and Mooney,

1983; Heal, 2000; Pimentel *et al.*, 2001; Kremen and Ostfeld, 2005; Millennium Ecosystem Assessment, 2005a). Nevertheless, in many cases, the costs of replacing services have been shown to exceed the costs of restoring or sustaining them, particularly in the case of water, and because policy makers often respond more to financial than to academic arguments, more integrative valuation approaches can generate positive policy shifts (TEEB, 2010).

It is not always the case that options are simple choices between meeting human needs through services from ecosystems or through their artificial replacement. Most landscapes are now highly managed, and built (physical) water infrastructure invariably coexists with natural (ecosystem) infrastructure, presenting increasing opportunities to manage both together to improve efficiency (see the example of the Itaipu watershed in Chapter 9).

Valuing ecosystem services can assist considerations of the costs and benefits of different options for achieving water and food security, and set the issues in their proper broader context. Table 3.1 provides an example of the valuation of ecosystem services delivered by various ecosystem types at the global scale. Although not necessarily applicable at the local scale, the results illustrate a number of important points. Collectively, values derived from regulatory, supporting and cultural services, generally outstrip values for provisioning services (goods produced) in all areas, and by a considerable margin (the value of ecosystem services in agriculture is further examined in Chapter 4). Despite this, most areas have historically been managed almost exclusively for provisioning services (in particular for food, which usually delivers among the lowest values of all). Water-related services (including water regulation and water-driven supporting services) generate some of the highest values of all.

Previously, Costanza *et al.* (1997) had suggested that wetlands provide more valuable food per hectare annually than other ecosystems: the total global value of food from wetlands was estimated at US$84.5 billion, while four times the area in cropland was calculated to produce a food value of US$75.6 billion. This was explained by the difference between high-value fish and shrimps, versus low-value grains. However, wetlands are the most valuable of all ecosystem types, not only because of aquatic food production but (in fact) primarily because they yield high benefits by providing and regulating water. Yet, despite this, wetlands show the most rapid rate of loss among all biomes – principally through agricultural impacts on water and the conversion of wetlands to farming (see also Chapter 7).

Hurricane Katrina, in 2005, prompted one of the most comprehensive and relevant detailed assessments of ecosystem services, as affected in particular by agriculture. This conclusively illustrated the pitfalls of sector-based planning for land and water resources management, and the economic and human costs of ignoring ecosystem services in river basin planning and development (Box 3.2).

Conclusions

The concept of 'sustainable food production' involves achieving the necessary increases in agricultural productivity, while simultaneously bringing the impacts of agriculture on ecosystems within manageable limits and in the face of significant resource challenges (as outlined in Chapters 1 and 2). The ecosystem setting of water within agroecosystems, and the way in which this determines the benefits (ecosystem services) that water provides, both within and beyond agriculture, offers a framework for identifying solutions to achieve sustainable agriculture. Further expansion of this approach is provided in subsequent chapters. To many readers, these concepts will not be new, but there is ample evidence that they are not being mainstreamed into agricultural planning and management. If they

Table 3.1. Estimation of the annual average value of ecosystem services of terrestrial biomes (in 2007 US$[a]/ha; adapted from van der Ploeg et al., 2010).

Ecosystem services	Tropical forests	Other forests	Woodland	Grass	Wetlands	Lakes and rivers
Provisioning						
Food production	121	496	68	54	709	94
Water supply	300	152		378	1,598	3,361
Other provisioning[b]	1,466	45	291	22	433	
Cultural	373	25	–[c]	4	3,218	1,337
Regulatory[d]						
Water flow regulation	19	1	–	–	4,660	–
Extremes	92	–	–	–	1,569	–
Other regulatory	1,711	143	432	686	1,460	2,642
Supporting[e]	1,008	399	–	99	2,104	–
Annual total (2007 US$/ha)[f]	5,088	1,261	792	1,244	15,752	7,433

[a] The international (Geary-Khamis) dollar is a hypothetical unit of currency that has the same purchasing power parity that the US$ had in the USA at a given point in time, in this case 2007, for which the unit is abbreviated 2007 US$.
[b] Other provisioning services include raw materials, and genetic, medicinal and ornamental resources.
[c] A nil return (–) indicates that insufficient data were available.
[d] Regulatory services include water flow regulation; waste treatment and water purification; moderation of extreme events such as floods, droughts and storms; and other regulatory services such as influence on air quality, climate regulation, erosion prevention, pollination and biological control.
[e] Supporting services include nutrient cycling, habitat services and maintenance of genetic resources.
[f] The total (van der Ploeg et al., 2010) may differ from calculated sum because of rounding.

Box 3.2. Agriculture and ecosystem services in the Mississippi River Delta (based on Batker *et al.*, 2010)

The history of management of the Mississippi River Basin, and impacts of this on its delta, present a case study illustrating the importance of valuing, and paying attention in agricultural policies to managing broader water-related ecosystem services.

River deltas are dynamic and complex ecosystems driven largely by hydrology, including the regular transfer of sediments and nutrients from the catchment into lowlands and the estuary. Their functioning underpins numerous ecosystem services, in particular land formation. This, in turn, delivers benefits through the maintenance of coastal stability and erosion regulation, thereby, for example, reducing disaster vulnerability. The Mississippi River Delta, in common with the deltas of many rivers, has been highly modified: its hydrology has been changed through water abstraction, principally for agriculture, while reservoir construction, also for hydropower, has interrupted sediment transfer. Additional physical infrastructure has had to be added, with high investment and operational costs; in effect, this is required to compensate for losses in the services originally provided by the ecosystem; examples include continual dyke and coastal defence development and maintenance in order to deal with a destabilizing estuary. The resulting degradation of the associated wetlands infrastructure is now widely regarded as a major contributing factor to the scale of economic and human losses resulting from hurricanes. In 2005, hurricane Katrina, in particular, was a catastrophic reminder of the pitfalls of paying insufficient attention to managing ecosystem services.

The study by Batker *et al.* (2010) estimated that if treated as an economic asset, the minimum asset value of the natural infrastructure provided by the Mississippi River Delta ecosystem would be US$330 billion to US$1.3 trillion (at 2007 values) in terms of hurricane and flood protection, water supply, water quality, recreation and fisheries. Importantly, the study also suggested that rehabilitation and restoration of this natural infrastructure would have an estimated net benefit of US$62 billion annually. This includes reduced disaster-risk vulnerability, and savings in capital and operational costs for physical infrastructure solutions (including factoring in the economic costs to existing users of reallocating water use).

As very pertinent to this book: agriculture has historically been a key driver of water-allocation policy in the Mississippi River (as in most river systems), yet the value of food and fibres produced by agriculture represents only a fraction of the value of the multitude of other services provided by the ecosystem, particularly by its wetlands (see Chapter 4 for more details on the values of ecosystem services in the Mississippi Delta).

This example illustrates: (i) the importance of understanding how ecosystems function and what ecosystem management offers in terms of cost-effective solutions (or avoiding problems in the first place); (ii) the importance of valuing ecosystem services more holistically; (iii) that the issue is not of one benefit versus another (in this case agriculture versus wetlands downstream) but of how to manage the river infrastructure (both physical and natural) to achieve the optimal outcome. Restoring optimal ecosystem services does not require agriculture to be compromised, but it does require a different risk management and investment approach.

There are now very many major cities, much larger than New Orleans, particularly in Asia, that are located in river deltas that are subject to a similar history of agriculture and water management. Hopefully, lessons learned from the Mississippi River Delta can help to achieve food security in these areas that is not at the expense of other security needs.

were, this book would not need to be written and sustainable agriculture would already have been achieved. This chapter identifies the opportunity to move beyond seeing the agriculture–environment relationship as one of conflict and trade-offs to looking at how improved ecosystem based management can deliver solutions that will simultaneously increase agricultural productivity and deliver improved broader ecosystem benefits.

References

Aylward, B., Bandyopadhyay, J. and Belausteguigotia, J.-C. (2005) Freshwater ecosystem services. In: Chopra, K., Leemans, R., Kumar, P. and Simons, H. (eds) *Ecosystems and Human Well-being: Policy Responses, Volume 3. Findings of the Responses Working Group of the Millennium Ecosystem Assessment.* Millennium Ecosystem Assessment and Island Press, Washington, DC, pp. 213–256.

Batker, D., de la Torre, I., Costanza, R., Swedeen, P., Day, J., Boumans, R. and Bagstad, K. (2010) *Gaining Ground. Wetlands, Hurricanes and the Economy: The Value of Restoring the Mississippi River Delta.* Earth Economics, Tacoma, Washington. Available at: http://www.eartheconomics.org/FileLibrary/file/Reports/Louisiana/Earth_Economics_Report_on_the_Mississippi_River_Delta_compressed.pdf (accessed February 2013).

Bennett, E.M., Peterson, G.D. and Gordon, L.J. (2009) Understanding relationships among multiple ecosystem services. *Ecology Letters* 12, 1394–1404. doi:10.1111/j.1461-0248.2009.01387.x

Bharucha, Z. and Pretty, J. (2010) The roles and values of wild foods in agricultural systems. *Philosophical Transactions of the Royal Society B* 365, 2913–2926. doi:10.1098/rstb.2010.0123

Cairns, J. Jr (1995) Editorial: Ecosystem services: an essential component of sustainable use. *Environmental Health Perspectives* 103 (6), 534.

Conway, G.R. (1987) The properties of agroecosystems. *Agricultural Systems* 24, 95–117. doi:10.1016/0308-521X(87)90056-4

Costanza, R. *et al.* (1997) The value of the world's ecosystem services and natural capital. *Nature* 387, 253–260. doi:10.1038/387253a0

Dale, V.H. and Polasky, S. (2007) Measure of the effects of agricultural practices on ecosystem services. *Ecological Economics* 64, 286–296. doi:10.1016/j.ecolecon.2007.05.009

Ehrlich, P. and Mooney, H. (1983) Extinction, substitution, and ecosystem services. *BioScience* 33, 248–254.

Ellis, E.C. (2011) Anthropogenic transformation of the terrestrial biosphere. *Philosophical Transactions of the Royal Society A* 369, 1010–1035. doi:10.1098/rsta.2010.0331

Ellis, E.C. and Ramankutty, N. (2008) Putting people in the map: anthropogenic biomes of the world. *Frontiers in Ecology and the Environment* 6, 439–447. doi:10.1890/070062

Falkenmark, M. (2003) Freshwater as shared between society and ecosystems: from divided approaches to integrated challenges. *Philosophical Transactions of the Royal Society B* 358, 2037–2049. doi:10.1098/rstb.2003.1386

Fresco, L.O. (2005) Water, food and ecosystems in Africa. *Spotlight Magazine*, 3 February 2005. Food and Agriculture Organization of the United Nations, Rome. Available at: www.fao.org/ag/magazine/0502sp1.htm (accessed December 2012).

Gandini, G. and Villa, E. (2003) Analysis of the cultural value of local livestock breeds: a methodology. *Journal of Animal Breeding and Genetics* 120, 1–11. doi:10.1046/j.1439-0388.2003.00365.x

Heal, G. (2000) Valuing ecosystem services. *Ecosystems* 3, 24–30. doi:10.1007/s100210000006

Kremen, C. (2005) Managing ecosystem services: what do we need to know about their ecology? *Ecology Letters* 8, 468–479. doi:10.1111/j.1461-0248.2005.00751.x

Kremen, C. and Ostfeld, R. (2005) Measuring, analyzing, and managing ecosystem services. *Frontiers in Ecology and the Environment* 3, 540–548. doi:10.1890/1540-9295(2005)003[0540:ACTEMA]2.0.CO;2

Mayers, J., Batchelor, C., Bond, I., Hope, R.A., Morrison, E. and Wheeler, B. (2009) *Water Ecosystem Services and Poverty Under Climate Change: Key Issues and Research Priorities. Report of a Scoping Exercise to Help Develop a Research Programme for the UK Department for International Development.* Natural Resource Issues 17, International Institute for Environment and Development, London. Available at: http://pubs.iied.org/pdfs/13549IIED.pdf (accessed December 2012).

McIntyre, B.D., Herren, H.R., Wakhungu, J. and Watson, R.T. (eds) (2008) *Agriculture at a Crossroads. International Assessment of Agricultural Knowledge, Science and Technology for Development (IAASTD): Global Report.* Island Press, Washington, DC. Available at: http://www.agassessment.org/reports/IAASTD/EN/Agriculture%20at%20a%20Crossroads_Global%20Report%20(English).pdf (accessed February 2013).

Millennium Ecosystem Assessment (2005a) *Ecosystems and Human Well-being: Synthesis. A Report of the Millennium Ecosystem Assessment.* World Resources Institute and Island Press, Washington, DC. Available at: www.maweb.org/documents/document.356.aspx.pdf (accessed February 2013).

Millennium Ecosystem Assessment (2005b) *Ecosystems and Human Well-being: Wetlands and Water – Synthesis. A Report of the Millennium Ecosystem Assessment.* World Resources Institute, Washington, DC. Available at: www.maweb.org/documents/document.358.aspx.pdf (accessed February 2013).

Molden, D. (ed.) (2007) *Water for Food, Water for Life: Comprehensive Assessment of Water Management in Agriculture*. Earthscan, London, in association with International Water Management Institute (IWMI), Colombo.

MRC (2003) Mekong River Awareness Kit. Interactive Self-Study CD-ROM. Mekong River Commission, Vientiane, Lao PDR/Phnom Penh, Cambodia.

OECD (2001) Glossary: Agroecosystem. In: *Environmental Indicators for Agriculture: Methods and Results, Volume 3*. Organisation for Economic Co-operation and Development, Paris, p. 399. Available at: http://www.oecd.org/tad/sustainableagriculture/40680869.pdf (accessed February 2013).

Pimentel, D., Gatto, M. and de Leo, G. (2001) Pricing biodiversity and ecosystem services. *BioScience* 51, 270–272.

TEEB (2010) *The Economics of Ecosystems and Biodiversity: Mainstreaming the Economics of Nature: A Synthesis of the Approach, Conclusions and Recommendations of TEEB*. Prepared by Sukhdev, P., Wittmer, H., Schröter-Schlaack, C., Nesshöver, C., Bishop, J., ten Brink, P., Gundimeda, H., Kumar, P. and Simmons, B. United Nations Environment Programme TEEB (The Economics of Ecosystems and Biodiversity) Office, Geneva, Switzerland. Available at: http://www.teebweb.org/wp-content/uploads/Study%20and%20Reports/Reports/Synthesis%20report/TEEB%20Synthesis%20Report%202010.pdf (accessed February 2013).

UCC-Water (2008) *Addressing Environmental Aspects of IWRM*. Concepts and Issues Paper No. 1, UNEP Collaborating Centre on Water and Environment, Hørsholm, Denmark.

UNEP (2009) *Ecosystem Management Programme. A New Approach to Sustainability*. United Nations Environment Programme, Nairobi. Available at: http://www.unep.org/ecosystemmanagement/Portals/7/Documents/EMP-Booklet.pdf (accessed February 2013).

UNFPA (2009) *State of World Population 2009. Facing a Changing World: Women, Population and Climate*. United Nations Population Fund, New York. Available at: www.unfpa.org/swp/2009/ (accessed December 2012).

van der Ploeg, S., de Groot, R.S. and Wang, Y. (2010) *The TEEB Valuation Database: Overview of Structure, Data and Results*. Foundation for Sustainable Development, Wageningen, the Netherlands.

Vergara, W. and Scholz, M. (eds) (2011) *Assessment of the Risk of Amazon Dieback*. World Bank, Washington, DC.

Wilson, M. and Carpenter, S. (1999) Economic valuation of freshwater ecosystem services in the United States: 1971–1997. *Ecological Applications* 9, 772–783. doi:10.1890/1051-0761(1999)009[0772:EVOFES]2.0.CO;2

Wood, S., Sebastian, K. and Scherr, S.J. (2000) *Pilot Analysis of Global Ecosystems: Agroecosystems. A Joint Study by International Food Policy Research Institute and World Resources Institute*. International Food Policy Research Institute and World Resources Institute, Washington, DC. Available at: www.wri.org/publication/pilot-analysis-global-ecosystems-agroecosystems (accessed January 2012).

WRI, UNDP, UNEP and World Bank (2008) *World Resources 2008: Roots of Resilience: Growing the Wealth of the Poor – Ownership – Capacity – Connection*. World Resources Institute in collaboration with United Nations Development Programme, United Nations Environment Programme and World Bank. World Resources Institute, Washington, DC.

Zhang, W., Ricketts, T.H., Kremen, C., Carney, K. and Swinton, S.M. (2007) Ecosystem services and disservices to agriculture. *Ecological Economics* 64, 253–260. doi:10.1016/j.ecolecon.2007.02.024

4 Challenges to Agroecosystem Management

Petina L. Pert,[1]* Eline Boelee,[2] Devra I. Jarvis,[3] David Coates,[4] Prem Bindraban,[5] Jennie Barron,[6] Rebecca E. Tharme[7] and Mario Herrero[8]

[1]Commonwealth Scientific and Industrial Research Organisation (CSIRO), Cairns, Queensland, Australia; [2]Water Health, Hollandsche Rading, the Netherlands; [3]Bioversity International, Rome, Italy; [4]Secretariat of the Convention on Biological Diversity (CBD), Montreal, Canada; [5]World Soil Information (ISRIC) and Plant Research International, Wageningen, the Netherlands; [6]Stockholm Environment Institute, University of York, UK and Stockholm Resilience Centre, Stockholm University, Stockholm, Sweden; [7]The Nature Conservancy (TNC), Buxton, UK; [8]Commonwealth Scientific and Industrial Research Organisation (CSIRO), St Lucia, Queensland, Australia

Abstract

As growth in population, gross domestic product (GDP) and consumption continues, further demands are placed on land, water and other resources. The resulting degradation can threaten the food security of poor people in fragile environments, particularly those whose livelihoods rely largely on agricultural activities. The concept of diversified or multifunctional agroecosystems is a relatively recent response to the decline in the quality of the natural resource base. Today, the question of agricultural production has evolved from a purely technical issue to a more complex one characterized by social, cultural, political and economic dimensions. Multifunctional agroecosystems carry out a variety of ecosystem services, such as the regulation of soil and water quality, carbon sequestration, support for biodiversity and sociocultural services, as well as meeting consumers' needs for food. In turn, these systems also rely on ecosystem services provided by adjacent natural ecosystems, including pollination, biological pest control, maintenance of soil structure and fertility, nutrient cycling and hydrological services. However, poor management practices in agroecosystems can also be the source of numerous disservices, including loss of wildlife habitat, nutrient runoff, sedimentation of waterways, greenhouse gas emissions, and pesticide poisoning of humans and non-target species. This chapter discusses the challenges to agroecosystem management, and how adopting a diversified approach will enable farmers to farm longer and more sustainably in an environment of greater uncertainty, in the face of climate change.

* E-mail: petina.pert@csiro.au

Background

The impacts of population growth and other demographic changes on ecosystems can vary over time. Population growth and urban sprawl will result in more people using more resources and placing more pressure on ecosystem services (see Chapter 2). Increasing populations require more habitable and arable land, which often results in the conversion of natural ecosystems and, ultimately, in the breakdown of ecosystems. There is increasingly negative feedback concerning the interactions between food security, agriculture, water and ecosystem services (Nellemann *et al.*, 2009). Food security is further threatened by reduced yields associated with depleted water quantity, reduced water quality, degradation of other natural resources (such as soil fertility) and the simplification of agricultural systems that have lost their inherent biotic components for regulating pest and disease infestations. Unsustainable agricultural practices can have profound, damaging side effects on livelihoods and ecosystem functioning, and in the long term could potentially depress or reverse productivity gains and increase poverty. At the same time, the availability of other natural resources (land, phosphorus and energy) is predicted to start running out by the end of this century (McIntyre *et al.*, 2008). Efforts to reactivate farmland, e.g. through the use of agrochemicals, have a substantial impact on other ecosystem functions. In turn, dysfunctional ecosystem services further affect the agroecosystems and their production systems.

The Millennium Ecosystem Assessment (MA) of 2005 suggested that in the next 50–100 years, major agricultural decisions would come in the form of trade-offs, especially 'between agricultural production and water quality, land use and biodiversity, water use and aquatic biodiversity, and current water use for agricultural production' (Nelson, 2005). Four scenarios and an adapted version of the MA framework were used in Australia to identify trade-offs between the ecosystem service of water regulation and stakeholders in the Great Barrier Reef's Tully–Murray Catchment (Butler *et al.*, 2011). While the most direct trade-off was found to be food and fibre production versus water quality regulation,

synergies were also identified with floodplain fisheries (Butler *et al.*, 2011).

As discussed in Chapter 3, greater understanding and appreciation of the role of the services provided by a variety of ecosystems, including agroecosystems, could assist in moving beyond 'trade-offs' to address the challenges of ecosystem management for long-term sustainable food production in many ways. The growing demands for food, coupled with land and water management practices that cause degradation and erode the natural resource base, place substantial constraints on the ecosystem services provided by and inherent within these agroecosystems (Abel *et al.*, 2003; Sandhu *et al.*, 2010).

Agriculture and ecosystem services are thus interrelated in at least four ways: (i) agro-ecosystems generate beneficial ecosystem services such as soil retention, food production and aesthetic benefits; (ii) agroecosystems receive beneficial ecosystem services from other ecosystems, such as pollination from non-agricultural ecosystems; (iii) ecosystem services from non-agricultural systems may be affected by agricultural practices; and finally (iv) the biological diversity within agricultural ecosystems provides regulating and supporting ecosystem services in addition to production services. For food security in the short term, provisioning services are crucial; however, for securing access to food for all in the future, and in the long term, regulatory and supporting services are just as important. The ecosystem services approach requires adaptive management, because its implementation depends on local, national or even global conditions.

Comparing the Economic Values of Ecosystem Services

Decisions on the management of agro-ecosystem services will typically involve balancing social, economic and environmental considerations, some of them among different services (Millennium Ecosystem Assessment, 2005; see Chapter 3). For example, managing a landscape to maximize food production will probably not maximize water purification for people downstream, and native habitats conserved near agricultural fields may provide both crop pollinators and crop pests (Steffan-Dewenter *et al.*, 2001). The question about whether intensive or extensive agriculture best optimizes the various trade-offs associated with the provision of ecosystem services is an important issue requiring targeted research.

Connections between ecological sustainability and human well-being can be expressed by using the concept of 'ecological character': the various components and processes in an ecosystem that underpin the delivery of ecosystem services (Millennium Ecosystem Assessment, 2005). Without managing for the sustainability of ecological character, the long-term ability of an ecosystem to support human well-being may be compromised. These kinds of management trade-offs often require decision makers to estimate the marginal values of ecosystem services, and to capture the costs and benefits of a specific quantity and quality of services (Daily, 1997) for men and women and different social groups.[1] Marginal value is used in this process because monetary valuation cannot express the overall importance of environmental goods and services (see Chapter 3), only the value of the resource if there were to be a little more or a little less of it (Heal, 2000). Therefore, the value of an ecosystem service reflects its availability. Water is a good example here: it is important and renewable but not replaceable. However, water is often provided freely or at a minimal cost to consumers. The price to consumers only pays for the cost of transmitting water (e.g. water treatment plants), which does not reflect the value of the water itself and gives no information on what consumers would be willing to pay if there were a little more or a little less of the resource (Heal, 2000).

Ecosystem services can also be used to compare different ecosystem types in terms of their contributions to the availability of a certain service. Most commonly, 'total valuation' is the tool used to bring environmental services into decision-making processes where trade-offs between conservation and development need to be comparatively assessed (Emerton, 2005). Total valuation attempts to account for all of the characteristics of an ecosystem; these include 'its resource stocks or assets, flows of environmental services, and the attributes of the ecosystem as a whole' (Millennium

Ecosystem Assessment, 2005). As mentioned above, this is an incomplete process that is limited in its capacity to value ecosystems fully, though as Daily (1997) points out, 'markets play a dominant role in patterns of human behavior, and the expression of value – even if imperfect – in a common currency helps to inform the decision-making process'.

For the quantification of the values of ecosystem services at the country level, a useful concept has been proposed by Dasgupta (2010), who argues that neither gross domestic product (GDP) nor the human development index (HDI) can determine whether development is sustainable. An assessment of *wealth per capita* is much more useful as it includes the total of all capital assets: infrastructure such as buildings and roads, health, skills, knowledge and institutions, and also natural capital, which may easily be left out of other assessments (Dasgupta, 2010).

These methods are increasingly important to today's decisions on agricultural water use. Bennett *et al.* (2005) point out that, with growing demands on food production and water use, demands on ecosystem services, in many cases, could surpass the capacity of certain ecosystems to supply these services. In these contexts, decision makers will need to draw a balance between the production of various services in ecosystems on the one hand, and the social and economic benefits and risks of using technology to provide them on the other (Bennett *et al.*, 2005). With a clear understanding of ecosystem services and their values, agroecosystems and non-agricultural terrestrial ecosystems can be compared (Power, 2010). Many goods and provisioning services come from non-agricultural land (such as food, fodder, fibre and timber), and in decisions over water allocation the whole range of ecosystem services, their benefits (values) and costs (social, financial, water) have to be taken into account (TEEB, 2010). Only then can well-balanced decisions be made about which ecosystem services are to be enhanced, at the expense of which other services, or about how ecosystems can be optimized to provide the widest range of ecosystem services (Power, 2010).

Finally, any successful decision making will depend on farmers and the farming community having the knowledge and leadership capacity to evaluate the benefits that any action will have for them (Jarvis *et al.*, 2011). This, in turn, will be dependent on systems that are in place that support activities taken by local, national and international organizations and agencies towards strengthening local institutions so as to enable farmers to take a greater role in the management of their resources.

Agroecosystems have an important role to play in food security but they have also been associated with negative impacts on other ecosystems. When compared with other groups of ecosystems, or biomes, the total value of ecosystem services from cropland is relatively low, even for food production alone. For example, in the Mississippi Delta, the total annual value of agricultural land ranged from US$195 to US$220/ha, of which US$85/ha was from food production; this level of production fell behind that of most other ecosystem types (including forests and, in particular, wetlands, where annual food production was valued at US$145 to US$3346/ha) (Batker *et al.*, 2010; and Box 3.2, Chapter 3).

In contrast, some other studies have found higher annual values for food production in cultivated systems: US$667/ha in South Africa, US$1516/ha in El Salvador, and as high as US$3842/ha and US$7425/ha in Israel (van der Ploeg *et al.*, 2010). However, it is not clear how this compares with the average values of other biomes as listed in Table 3.1 (Chapter 3). There have been very few studies that have attempted to value ecosystem services in agriculture, even though assessments indicate that the value to agriculture is enormous (Power, 2010), and various estimates do suggest a real underestimation of the benefits of non-agricultural ecosystems for food production and possibly for food security. In a study in Denmark, Porter *et al.* (2009) estimated via field-scale ecological monitoring and economic value-transfer methods, the market and non-market ecosystem service value of a combined food and energy (CFE) agroecosystem that simultaneously produces food, fodder and bioenergy.

Discrepancies in estimations of the economic values of ecosystem services occur, in part, because land and water use planning are based on limited sector-based considerations, which do not factor in the overall values of all services that any ecosystem delivers. Hence, agricultural land has such a low value in terms of output because it tends to be managed for a single service (food production), often with significant negative consequences on other services (e.g. through pollution). Another reason might be that the value of food production is measured in terms of market prices, whereas the value of other ecosystem services reflects avoided societal costs that are normally much higher but for which there are no marketplaces (with the exception of carbon). Nevertheless, food production will always remain a priority and does not necessarily have to come at the expense of other services (Bennett *et al.*, 2009; Keys *et al.*, 2012). Cases exist in which investments in sustainable agriculture have generated co-benefits in raising food production, while at the same time improving ecosystem services and functions (Pretty *et al.*, 2006; see examples in other chapters).

Understanding Agroecosystem Services

Managing agricultural land to deliver multiple services considerably improves the value of the land. However, in order to enhance improved services – such as carbon storage, erosion control, water retention, waste treatment, regulation of pests and diseases, and cultural and recreational values including tourism – their values must be understood in comparison with agricultural income. Ideally, these added services would not conflict with agricultural production in many cases but rather improve both its productivity and its sustainability, with beneficial impacts on surrounding ecosystems as well (see Chapter 9 for more information on managing a wider range of agroecosystem services).

Over the years, agricultural systems have evolved into diverse agroecosystems, some of which are rich in biodiversity and provide ecosystem services in addition to food production. Examples are wet rice–poultry farming systems and the practice of increased diversity of crop varieties within farmers' fields, which has been shown to reduce the risk of crop loss to pest diseases (Jarvis *et al.*, 2007; Mulumba *et al.*, 2012).

Water management in agroecosystems can create competition with wider environmental requirements and affect water flow downstream. Decisions on water use require mechanisms in which the needs of both the farmers and the ecosystem services are met, e.g. by buying irrigation water from farmers to sustain or rehabilitate ecosystems and their services (Molden and de Fraiture, 2004). These decisions need a broader consideration of ecosystem services in agroecosystems. This consideration should take into account which services are enhanced at the expense of which other ecosystem services, and which services benefit mostly poor women, men and other vulnerable groups. In agroecosystems, food production is again underpinned by a reliable availability of water. Tools, such as the polyscape tool, are being developed that allow the quantification of trade-offs and synergies among the impacts of water- and land-use interventions on different ecosystem services (Box 4.1).

Box 4.1. Polyscape tool for comparing impacts on ecosystem services.

One of the new tools under development for assessing ecosystem services is the polyscape tool (adapted from Jackson *et al.*, 2013). This allows the quantification of trade-offs and synergies among the impacts of land-use interventions such as the changing of tree cover. Small catchment maps indicate with colours where, for example, new tree cover would be most desirable to enhance woodland habitat connectivity, reduce flow accumulation, have minimum impact on farm productivity and reduce sediment transport (Fig. 4.1). When the four benefits are traded off in the large map, there is only a small area of the catchment where tree placement benefits all goals. To substantially enhance some ecosystem services by increasing tree cover, farmers would need to be well compensated for loss of production; for other ecosystem services, only certain farms in the landscape would be important, i.e. different bits of the landscape would have different values for each service considered.

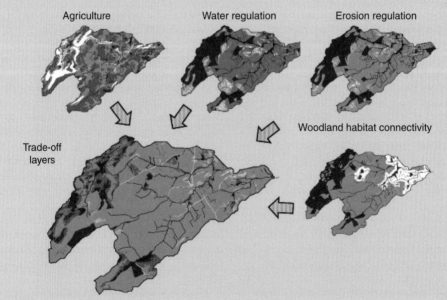

Fig. 4.1. Example of the application of the polyscape tool (figure components provided by Tim Pagella) to explore trade-offs and synergies of the impacts of tree cover on ecosystem services. In the four individual maps, darker areas represent high value for the service and lighter areas opportunities for improvement. In the combined map, darker areas represent trade-offs (where improvements in one service could be at the expense of others), whilst lighter areas mean that changes will provide multiple benefits (synergies).

Agriculture is thus faced with significant challenges regarding water use and availability. Solutions, which are based largely on the more efficient use of water in agriculture, do exist (see Chapters 5 and 8), but agriculture can also be managed differently, in such a way as to enhance ecosystem services and increase the capacities of low-income male and female farmers (Molden, 2007; see also Chapter 9). This change in thinking, and in the way that agroecosystems are managed, is crucial for global food security. The major challenge then lies in quantifying values and measuring feedback cycles (Nicholson *et al.*, 2009; Taffetani *et al.*, 2011), and more research is required into ecosystem services, especially those associated with water (Carpenter *et al.*, 2009) and with other components of the agricultural production systems.

Ecosystem services and fish

Important fisheries that depend on healthy aquatic ecosystems are endangered. Because fish provide 21% of animal protein in Africa, and 28% in Asia (World Commission on Dams, 2000), a loss of fisheries can be detrimental for food security. The link with management of inland aquatic ecosystems is clear, as almost 50% of global fish consumption comes from aquaculture, and in Africa almost half is from inland fisheries (UNEP, 2010). In order to avoid further degradation, fundamental changes are required to establish an ecosystem-based catchment management approach (IUCN, 2000).

Faced with declining wild fish stocks, over-exploitation of target species and by-catch of other species, the fishing industry is giving way to aquaculture, which is reported to be the world's fastest growing food sector – at an average growth rate of 6.8%/year (Medialdea, 2010). In 2006, it was reported that 53 million t of fish (or half of all fish consumed in the world) were produced by the aquaculture industries (Medialdea, 2010). At the same time, fisheries are increasingly less 'wild', as stock enhancement and the establishment of culture-based fisheries are increasingly viewed as potential means of bolstering catches. None the less, the potential negative ecological and social impacts of such practices demand comprehensive and rigorous assessment, with appropriate mitigation and control measures, before they are implemented. For example, antibiotics and other chemicals used in fish farms can seep into surrounding waters, and sensitive coastal areas and wetlands are also disrupted or destroyed in the development of the industry. Additionally, aquaculture appropriates a range of environmental goods and services that may lead to adverse environmental impacts, and affect the ability of stocks and flows of ecosystem services to sustain other productive activities, which could again result in disputes and conflicts.

Ecosystem impacts of livestock production

Livestock systems occupy about 30% (Steinfeld *et al.*, 2006) to 45% (Herrero *et al.*, 2010) of the planet's ice-free terrestrial surface area. This makes livestock the single largest agricultural use of land globally, either directly through grazing or indirectly through the consumption of fodder and feed grains. Livestock is also a significant global asset, with a value of at least US$1.4 trillion in the least developed countries, excluding the value of infrastructure and land (Herrero *et al.*, 2010). The accelerating demand for livestock products (see Box 2.2, Chapter 2) is increasingly being met by intensive (industrialized) production systems, especially for chickens and pigs in Asia (Thornton, 2010). Thus, between 1995 and 2005, bovine and ovine meat production increased by about 40%, pig meat production rose by nearly 60% and poultry meat production doubled (Steinfeld *et al.*, 2006). Livestock production has important implications for ecosystem services, with environmental impacts on water scarcity, nutrient cycling, climate change and land degradation, as well as human impacts such as public health and the exclusion of smallholder producers.

Livestock production emits large amounts of greenhouse gases (Box 2.3, Chapter 2). However, the mitigation potential in the livestock sector is very large (1.74 Gt CO_2 eq./ year; see Smith *et al.*, 2008; World Bank, 2009), with improved feeding practices, manure and land use management practices representing over 80% of this potential (Smith et al., 2008; Chapter 9).

A well-known linkage between livestock and soil productivity is the cycling of biomass (natural vegetation, crop residues) through animals (cattle, sheep, goats) into excreta (manure, urine) that fertilize the soil. Globally, manure contributes 14% of nitrogen, 25% of phosphorus and 40% of potassium nutrient inputs to agricultural soils (Bouwman et al., 2011). The types and amounts of manure nutrients available for recycling are highly influenced by differences in land use and in the spatial and temporal distribution of livestock as dictated by animal management, and also by seasonal differences in animal diet. When not carefully managed, nutrient surpluses from livestock waste and fertilizer used for feed production may result in eutrophication of surface waters and groundwater contamination in places where large animals congregate, such as in industrial peri-urban systems. These ecosystem disservices posed by water contamination from livestock excreta and dung residues can cause health hazards (Herrero et al., 2009).

Then again, there are numerous potential situations where co-benefits emerge between livestock production and the maintenance of ecosystem services (see Chapter 9). These examples may not be readily available, as they require in-depth analysis of scientific as well as indigenous evidence, and therefore come at a (knowledge-intensive) cost. Herrero et al. (2009) formulated useful guiding questions on livestock, ecosystems and livelihoods to help to identify knowledge gaps. In analysing environmental impacts and ecosystem services, it is important to distinguish between extensive and intensive livestock production. Although livestock grazing is the largest user of land globally, most of the world's animal production comes from intensive industrialized production in developed countries, closely followed by rainfed mixed crop–livestock systems in developing countries. These intensively farmed areas are the focal points for ecosystem degradation. For example, in Ethiopia, 45% of the estimated soil loss occurs from the 13% of the country under cultivation, but grazing lands, which cover about half of the country, account for only 21% of the soil loss (Hurni,

1990). Some livestock herding systems in Africa have managed large areas in a semi-natural state, maintaining vegetation cover and indirectly preserving vital ecosystem services.

Sustainable growth and intensification of livestock production systems will be required to cater for increasing demands for livestock products, while mitigating the negative effects of the sector (Tarawali et al., 2011). Substantive investments and policies are essential to implement the measures above (World Bank, 2009). With more sustainable livestock production systems, the increased demands for animal products could be satisfied at the same time as maintaining environmental flows and services.

Land degradation and erosion

Soil degradation, such as by water or wind erosion, compaction, salinization, nutrient depletion and fertility decline, physical deterioration, contamination and sealing, is considered to be a main cause of hampering growth in agricultural productivity (Sanchez et al., 1997). The impact of soil degradation on yields in China was estimated as a reduction in food production capacity on the current arable land area from 482 Mt in 2005 to 412 Mt by 2050, with the same relative yield loss projected in the next 15 years as in the past 15 years (Bindraban et al., 2012), though such estimates do not account for underlying processes; hence, for identifying viable solutions, more detailed studies at a lower level will be required. In addition to physical factors, land degradation has many social roots, including lack of land tenure, careless extractivism, indifferent or corrupt governments, lack of access to finance and resources, population pressure and a dearth of educational opportunities.

In many parts of the world, land degradation has increased over the past two decades, mostly as a result of poor land management, including uncontrolled soil erosion, overgrazing, and the limited application and availability of appropriate types of fertilizers. In sub-Saharan Africa, more than 40% of the land is threatened by land degradation (Vlek et al., 2010). Loss of organic matter, e.g. through entire crop removal, and the physical degradation of soil not only reduce nutrient availability but also result in lower water infiltration rates and porosity, and these may affect the resilience of agroecosystems, local and regional water productivity, and even global carbon cycles. Accelerated on-farm soil erosion leads to substantial yield losses and contributes to downstream sedimentation, which can degrade natural water bodies and fill up water storage reservoirs and irrigation infrastructure (Vlek *et al.*, 2010; Bouma *et al.*, 2011).

The occurrence of land degradation is thus linked with low water productivity and impaired ecosystem services (Bossio *et al.*, 2008), and is often associated with high population pressure; nevertheless, its extent and its causative mechanisms are highly site specific (Muchena *et al.*, 2005). One way of dealing with this is to facilitate outmigration of people from vulnerable areas through the provision of education and credit services offering alternative livelihoods (World Bank, 2009). However, high population pressure and market demand can in itself trigger investments in labour-intensive conservation practices and natural resources management (Nelson, 2005).

Another argument for taking a landscape approach (more on that in Chapter 11), is the role of trees. Recent assessments suggest that almost half of all agricultural land has more than 10% tree cover, indicating that trees are a mainstream component of agricultural landscapes (Zomer *et al.*, 2009) and may provide forest functions to some extent. Tree cover in farming landscapes can have a large impact on the infiltration and penetration of water and, thereby, on catchment hydrology (Carroll *et al.*, 2004; Fig. 3.1, Chapter 3). Furthermore, when tree cover is changed, other ecosystem services besides water flow may also be affected, such as pollination and carbon storage, and these can also influence agricultural productivity (Harvey *et al.* 2006). The impact of changing tree cover on various ecosystem services depends on its amount, spatial configuration, species composition and management, so there is a need to get beyond generalizations and look at tree cover at the landscape scale in order to meet specific objectives, including the consideration of trade-offs and synergies among the ecosystem services affected (Jackson *et al.*, 2013).

Conclusions

In recent years, there has been an inexorable rise in the demand for food and for water to grow food. Particularly, the high demand for water and land in commercial farming systems and, with it, the increased risks of pollution have led to the need for more economically, socially and environmentally viable agricultural systems in order to avoid ecosystem destruction. This chapter has explored these demands and challenges within an agroecosystems management context.

Growing concerns about the negative changes produced by agriculture on various ecosystems across the world (key 'disservices' from agriculture) have been analysed. The Millennium Ecosystem Assessment showed that agriculture has dramatically increased its ecological footprint, not only in terms of negative impacts but also in terms of its supply of ecosystem services for rural communities. A discussion of the value of ecosystem services has provided a better understanding of the linkages between agriculture and ecosystem services, paving the road for management options that are addressed in subsequent chapters.

Note

[1] Differentiating the groups here is important because different groups, for example men and women, young and old, or poor and rich, make very different use of the services available to them and may value these services very differently. The different use various social groups make of water and ecosystems, and the impacts of that in relation to development and conservation projects, are discussed in more detail in other publications (e.g. Thompson and Swatuk, 2000; Goma Lemba *et al.*, 2001; Sudarshan, 2001; Hassan *et al.*, 2005; www.genderandwater.org).

References

Abel, N., Cork, S., Gorddard, R., Langridge, J., Langston, A., Plant, R., Proctor, W., Ryan, P., Shelton, D., Walker, B. and Yialeloglou, M. (2003) *Natural Values: Exploring Options for Enhancing Ecosystem Services in the Goulburn Broken Catchment.* CSIRO (Commonwealth Scientific and Industrial Research Organisation) Sustainable Ecosystems [now CSIRO Ecosystem Sciences], Canberra. Available at: www.ecosystemservicesproject.org/html/publications/docs/nveo/Natural_Values.pdf (accessed December 2012).

Batker, D., de la Torre, I., Costanza, R., Swedeen, P., Day, J., Boumans, R. and Bagstad, K. (2010) *Gaining Ground. Wetlands, Hurricanes and the Economy: The Value of Restoring the Mississippi River Delta.* Earth Economics, Tacoma, Washington. Available at: http://www.eartheconomics.org/FileLibrary/file/Reports/Louisiana/Earth_Economics_Report_on_the_Mississippi_River_Delta_compressed.pdf (accessed February 2013).

Bennett, E.M., Peterson, G.D. and Levitt, E.A. (2005) Looking to the future of ecosystem services. *Ecosystems* 8, 125–132. doi:10.1007/s10021-004-0078-y

Bennett, E.M., Peterson, G.D. and Gordon, L.J. (2009) Understanding relationships among multiple ecosystem services. *Ecology Letters* 12, 1394–1404. doi:10.1111/j.1461-0248.2009.01387.x

Bindraban, P.S. *et al.* (2012) Assessing the impact of soil degradation on food production. *Current Opinion in Environmental Sustainability* 4, 478–488.

Bossio, D., Noble A., Molden D. and Nangia V. (2008) Land degradation and water productivity in agricultural landscapes. In: Bossio, D. and Geheb, K. (eds) (2008) *Conserving Land, Protecting Water.* Comprehensive Assessment of Water Management in Agriculture Series 6. CAB International, Wallingford, UK in association with CGIAR Challenge Program on Water and Food, Colombo and International Water Management Institute (IWMI), Colombo, pp. 20–32.

Bouma, J., Droogers, P., Sonneveld, M.P.W., Ritsema, C.J., Hunink, J.E., Immerzeel, W.W. and Kauffman, S. (2011) Hydropedological insights when considering catchment classification. *Hydrology and Earth System Sciences* 15, 1909–1919.

Bouwman, L., Goldewijk, K.K., Van Der Hoek, K.W., Beusen, A.H.W., Van Vuuren, D.P., Willems, J., Rufino, M.C. and Stehfest, E. (2011) Exploring global changes in nitrogen and phosphorus cycles in agriculture induced by livestock production over the 1900–2050 period. *Proceedings of the National Academy of Sciences of the United States of America.* doi 10.1073/pnas.1012878108; correction (2012) doi:10.1073/pnas.1206191109.

Butler, J.R.A., Wong, G.Y., Metcalfe, D.J., Honzák, M., Pert, P.L., Rao, N., Van Grieken, M.E., Lawson, T., Bruce, C., Kroon, F.J. and Brodie, J.E. (2011) An analysis of trade-offs between multiple ecosystem services and stakeholders linked to land use and water quality management in the Great Barrier Reef, Australia. *Agriculture, Ecosystems and Environment.* doi:10.1016/j.agee.2011.08.017

Carpenter, S.R. *et al.* (2009) Science for managing ecosystem services: beyond the Millennium Ecosystem Assessment. *Proceedings of the National Academy of Sciences of the United States of America* 106, 1305–1312. doi:10.1073/pnas.0808772106

Carroll, Z.L., Bird, S.B., Emmett, B.A., Reynolds, B. and Sinclair, F.L. (2004) Can tree shelterbelts on agricultural land reduce flood risk? *Soil Use and Management* 20, 357–359. doi:10.1111/j.1475-2743.2004.tb00381.x

Daily, G. (1997) *Nature's Services: Societal Dependence on Natural Ecosystems.* Island Press, Washington, DC.

Dasgupta, P. (2010) Nature's role in sustaining economic development. *Philosophical Transactions of the Royal Society B* 365, 5–11. doi:10.1098/rstb.2009.0231

Emerton, L. (ed.) (2005) *Values and Rewards: Counting and Capturing Ecosystem Water Services for Sustainable Development.* IUCN Water, Nature and Economics Technical Paper No. 1, IUCN – The World Conservation Union, Ecosystems and Livelihoods Group Asia, Colombo. Available at: http://cmsdata.iucn.org/downloads/2005_047.pdf (accessed December 2012).

Goma Lemba, S., Nwukor, E. and Dagnachew, S. (2001) *Assessing Women and Environment. 5 Years after Beijing: What Efforts in Favour of African Women?* Economic Commission for Africa, Addis Ababa. Available at: http://www.uneca.org/acgs/Publications/en_0109_environment.pdf (accessed December 2012).

Harvey, C.A., Medina, A., Sanchez, D.M., Vilchez, S., Hernandez, B., Saenz, J.C., Maes, J.M., Casanoves, F. and Sinclair, F.L. (2006) Patterns of animal diversity in different forms of tree cover in agricultural landscapes. *Ecological Applications* 16, 1986–1999. doi:10.1890/1051-0761(2006)016[1986:POADID]2.0.CO;2

Hassan, R., Scholes, R. and Ash, N. (eds) (2005) *Ecosystems and Human Well-being: Current State and Trends, Volume 1. Findings of the Condition and Trends Working Group of the Millennium Ecosystem Assessment.* World Resources Institute and Island Press, Washington, DC.

Heal, G. (2000) Valuing ecosystem services. *Ecosystems* 3, 24–30. doi:10.1007/s100210000006

Herrero, M., Thornton, P.K., Gerber, P. and Reid, R.S. (2009) Livestock, livelihoods and the environment: understanding the trade-offs. *Current Opinion in Environmental Sustainability* 1, 111–120. doi:10.1016/j. cosust.2009.10.003

Herrero, M. *et al.* (2010) Smart investments in sustainable food production: revisiting mixed crop–livestock systems. *Science* 327, 822–825. doi:10.1126/science.1183725

Hurni H. (1990) Degradation and conservation of soil resources in the Ethiopian Highlands. In: Messerli, B. and Hurni, H. (eds) *African Mountains and Highlands: Problems and Perspectives.* African Mountains Association (AMA), Marceline, Missouri, pp. 51–63.

IUCN (2000) *Vision for Water and Nature: A World Strategy for Conservation and Sustainable Management of Water Resources in the 21st Century.* IUCN – The World Conservation Union, Gland, Switzerland.

Jackson, B., Pagella, T., Sinclair, F., Orellana, B., Henshaw, A., Reynolds, B., Mcintyre, N., Wheater, H. and Eycott, A. (2013) Polyscape: A GIS mapping framework providing efficient and spatially explicit landscape-scale valuation of multiple ecosystem services. *Landscape and Urban Planning* 112, 74–88. doi:10.1016/ j.landurbplan.2012.12.014

Jarvis, D.I., Brown, A.D.H., Imbruce, V., Ochoa, J., Sadiki, M., Karamura, E., Trutmann, P. and Finckh, M.R. (2007) Managing crop disease in traditional agroecosystems: the benefits and hazards of genetic diversity. In: Jarvis, D.I., Padoch, C. and Cooper, H.D. (eds) *Managing Biodiversity in Agricultural Ecosystems.* Published for Bioversity International, Rome by Columbia University Press, New York, pp. 292–319.

Jarvis, D.I., Hodgkin, T., Sthapit, B.R., Fadda, C. and Lopez-Noriega, I. (2011) An heuristic framework for identifying multiple ways of supporting the conservation and use of traditional crop varieties within the agricultural production system. *Critical Reviews in Plant Sciences* 30, 125–176. doi:10.1080/07352689.2 011.554358

Keys, P., Barron, J. and Lannerstad, M. (2012) *Releasing the Pressure: Water Resource Efficiencies and Gains for Ecosystem Services.* United Nations Environment Programme (UNEP), Nairobi and Stockholm Environment Institute, Stockholm.

McIntyre, B.D., Herren, H.R., Wakhungu, J. and Watson, R.T. (eds) (2008) *Agriculture at a Crossroads. International Assessment of Agricultural Knowledge, Science and Technology for Development (IAASTD): Global Report.* Island Press, Washington, DC. Available at: http://www.agassessment.org/reports/IAASTD/ EN/Agriculture%20at%20a%20Crossroads_Global%20Report%20(English).pdf (accessed February 2013).

Medialdea, J.M. (2010) A new approach to sustainable agriculture. *Solutions Journal,* June 2010. Available at: http://www.thesolutionsjournal.com/node/639 (accessed February 2013).

Millennium Ecosystem Assessment (2005) *Ecosystems and Human Well-being: Synthesis. A Report of the Millennium Ecosystem Assessment.* World Resources Institute and Island Press, Washington, DC. Available at: www.maweb.org/documents/document.356.aspx.pdf (accessed February 2013).

Molden, D. (ed.) (2007) *Water for Food, Water for Life: Comprehensive Assessment of Water Management in Agriculture.* Earthscan, London, in association with International Water Management Institute (IWMI), Colombo.

Molden, D. and de Fraiture, C. (2004) *Investing in Water for Food, Ecosystems and Livelihoods.* Blue Paper, Stockholm 2004, Discussion Draft. Comprehensive Assessment of Water Management in Agriculture, International Water Management Institute, Colombo. Available at: http://intranet.catie.ac.cr/intranet/ posgrado/CambioGlobal/19%20August/BluePaper%20water.pdf (accessed February 2013).

Muchena, F.N., Onduru, D.D., Gachin, G.N. and de Jager, A. (2005) Turning the tides of soil degradation in Africa: capturing the reality and exploring opportunities. *Land Use Policy* 22, 23–31. doi:10.1016/j. landusepol.2003.07.001

Mulumba, J.W., Nankya, R., Adokorach, J., Kiwuka, C., Fadda, C., De Santis, P. and Jarvis, P.I. (2012) A risk-minimizing argument for traditional crop varietal diversity use to recue pest and disease damage in agricultural ecosystems of Uganda. *Agriculture, Ecosystems and the Environment* 157, 70–86. doi:10.1016/j.agee.2012.02.012

Nellemann, C., MacDevette, M., Manders, T., Eickhout, B., Svihus, B., Prins, A.G. and Kaltenborn, B.P. (eds) (2009) *The Environmental Food Crisis – The Environment's Role in Averting Future Food Crises. A UNEP Rapid Response Assessment.* United Nations Environment Programme, GRID-Arendal, Norway. Available at: http://www.grida.no/files/publications/FoodCrisis_lores.pdf (accessed February 2013).

Nelson, G.C. (2005) Drivers of ecosystem change: summary chapter. In: Hassan, R., Scholes, R. and Ash, N. (eds) (2005) *Ecosystems and Human Well-being: Current State and Trends, Volume 1. Findings of the Condition and Trends Working Group of the Millennium Ecosystem Assessment.* World Resources Institute and Island Press, Washington, DC, pp. 73–76.

Nicholson, E. *et al.* (2009) Priority research areas for ecosystem services in a changing world. *Journal of Applied Ecology* 46, 1139–1144. doi:10.1111/j.1365-2664.2009.01716.x

Porter, J., Costanza, R., Sandhu, H., Sigsgaard, L. and Wratten, S. (2009) The value of producing food, energy, and ecosystem services within an agro-ecosystem. *Ambio* 38, 186–193.

Power, A.G. (2010) Ecosystem services and agriculture: tradeoffs and synergies. *Philosophical Transactions of the Royal Society B* 365, 2959–2971. doi:10.1098/rstb.2010.0143

Pretty, J., Noble, A.D., Bossio, D., Dixon, J., Hine, R.E., Penning de Vries, F.W.T. and Morison, J.I.L. (2006) Resource-conserving agriculture increases yields in developing countries. *Environmental Science and Technology* 40, 1114–1119. doi:10.1021/es051670d

Sanchez, P.A., Shepherd, K.D., Soule, M.J., Place, F.M., Buresh, R.J., Izac, A.-M.N., Mokwunye, A.U., Kwesiga, F.R., Ndiritu, C.G. and Woomer, P.L. (1997) Soil fertility replenishment in Africa: an investment in natural resource capital. In: Buresh, R.J., Sanchez, P.A. and Calhoun, F. (eds) *Replenishing Soil Fertility in Africa*. SSSA Special Publication 51, Soil Science Society of America, Madison, Wisconsin in association with International Centre for Research on Agroforestry (now World Agroforestry Centre; ICRAF), Nairobi, pp. 1–46.

Sandhu, H.S., Wratten, S.D. and Cullen, R. (2010) The role of supporting ecosystem services in arable farmland. *Ecological Complexity* 7, 302–310. doi:10.1016/j.ecocom.2010.04.006

Smith, P. *et al.* (2008) Greenhouse gas mitigation in agriculture. *Philosophical Transactions of the Royal Society B* 363, 789–813.

Steffan-Dewenter, I., Munzenberg, U. and Tscharntke, T. (2001) Pollination, seed set and seed predation on a landscape scale. *Proceedings of the Royal Society of London B* 268, 1685–1690. doi:10.1098/rspb.2001.1737

Steinfeld, H., Gerber, P., Wassenaar, T., Castel, V., Rosales, M. and de Haan, C. (2006) *Livestock's Long Shadow: Environmental Issues and Options*. Food and Agriculture Organization (FAO), Rome. Available at: www.fao.org/docrep/010/a0701e/a0701e00.HTM (accessed December 2012).

Sudarshan, R.M. (2001) Managing ecosystems for women's health and sustainable development. An illustration from the Kumaon, Uttaranchal. *Indian Journal of Gender Studies* 8, 207–222. doi:10.1177/097152150100800204

Taffetani, F., Rismondo, M. and Lancioni, A. (2011) Chapter 15. Environmental evaluation and monitoring of agro-ecosystems biodiversity. In: Grillo, O. and Venora, G. (eds) *Ecosystems Biodiversity*. InTech, Rijeka, Croatia, pp. 333–370. Available at: http://www.intechopen.com/books/ecosystems-biodiversity/environmental-evaluation-and-monitoring-of-agro-ecosystems-biodiversity (accessed February 2013). doi:10.5772/23366

Tarawali, S., Herrero, M., Descheemaeker, K., Grings, E. and Blümmel, M. (2011) Pathways for sustainable development of mixed crop livestock systems: taking a livestock and pro-poor approach. *Livestock Science* 139, 11–21. doi:10.1016/j.livsci.2011.03.003

TEEB (2010) *The Economics of Ecosystems and Biodiversity: Mainstreaming the Economics of Nature: A Synthesis of the Approach, Conclusions and Recommendations of TEEB*. Prepared by Sukhdev, P., Wittmer, H., Schröter-Schlaack, C., Nesshöver, C., Bishop, J., ten Brink, P., Gundimeda, H., Kumar, P. and Simmons, B. United Nations Environment Programme TEEB (The Economics of Ecosystems and Biodiversity) Office, Geneva, Switzerland. Available at: http://www.teebweb.org/wp-content/uploads/Study%20and%20Reports/Reports/Synthesis%20report/TEEB%20Synthesis%20Report%202010.pdf (accessed February 2013).

Thompson, L. and Swatuk, L. (2000) Gender and ecosystems: reframing Southern African "security". University of the Western Cape, Bellville, South Africa.

Thornton, P.K. (2010) Livestock production: recent trends, future prospects. *Philosophical Transactions of the Royal Society B* 365, 2853–2867. doi:10.1098/rstb.2010.0134

UNEP (2010) *Blue Harvest: Inland Fisheries as an Ecosystem Service*. WorldFish, Penang, Malaysia and United Nations Environment Programme, Nairobi.

van der Ploeg, S., de Groot, R.S. and Wang, Y. (2010) *The TEEB Valuation Database: Overview of Structure, Data and Results*. Foundation for Sustainable Development, Wageningen, the Netherlands.

Vlek, P.L.G., Le, Q.B. and Tamene, L. (2010) Assessment of land degradation, its possible causes and threat to food security in sub-Saharan Africa. In: Lal, R. and Stewart, B.A. (eds) *Food Security and Soil Quality*. Advances in Soil Science Series, CRC Press, Boca Raton, Florida, pp. 57–86.

World Bank (2009) *Minding the Stock: Bringing Public Policy to Bear on Livestock Sector Development*. Report No. 44110-GLB, World Bank, Agriculture and Rural Development Department, Washington, DC.

World Commission on Dams (2000) *Dams and Development: A New Framework for Decision-making*. Earthscan Publications, London and Sterling, Virginia.

Zomer, R.J., Trabucco, A., Coe, R. and Place, F. (2009) *Trees on Farm: Analysis of Global Extent and Geographical Patterns of Agroforestry*. ICRAF Working Paper No. 89, World Agroforestry Centre (ICRAF), Nairobi. Available at: www.worldagroforestrycentre.org/downloads/publications/PDFs/WP16263.PDF (accessed June 2011).

5 Water Use in Agroecosystems

Renate Fleiner,[1]* Delia Grace,[2] Petina L. Pert,[3] Prem Bindraban,[4] Rebecca E. Tharme,[5] Eline Boelee,[6] Gareth J. Lloyd,[7] Louise Korsgaard,[7] Nishadi Eriyagama[8] and David Molden[1]

[1]*International Centre for Integrated Mountain Development (ICIMOD), Kathmandu, Nepal;* [2]*International Livestock Research Institute (ILRI), Nairobi, Kenya;* [3]*Commonwealth Scientific and Industrial Research Organisation (CSIRO), Cairns, Queensland, Australia;* [4]*World Soil Information (ISRIC) and Plant Research International, Wageningen, the Netherlands;* [5]*The Nature Conservancy (TNC), Buxton, UK;* [6]*Water Health, Hollandsche Rading, the Netherlands;* [7]*UNEP–DHI Centre for Water and Environment, Hørsholm, Denmark;* [8]*International Water Management Institute (IWMI), Colombo, Sri Lanka*

Abstract

The integrated role of water in ecosystems and, in particular, in agroecosystems, as well as the multiple uses of water – across various sectors that have increasing demands, have been widely recognized. But regions and institutions are still struggling to resolve issues around water – be it scarcity, accessibility or degradation. Mostly, they are caught in conventional institutional and policy frameworks that have been set up based more on sectoral than on cross-sectoral principles, thus preventing them from achieving the ultimate goal of sustainability. This chapter analyses the current and future challenges related to water availability and water use for agriculture from this perspective. It looks at water quantity and quality, water infrastructure, and related governance and institutional aspects, using case studies from basins in different geographic regions.

Background

Agriculture uses about 70–72% of the total water that is withdrawn from surface and groundwater around the world (Molden, 2007; Wisser et al., 2008), and as much as about 90% in developing countries (Cai and Rosegrant, 2002). Shortages of water, and the means by which they can have a major effect on food production, are discussed by Strzepek and Boehlert (2010). It is estimated that food production needs to increase by at least 50%, and probably almost double, by 2050 in order to meet the needs of a growing population and changing consumer preferences for more water-intensive crops (Molden, 2007). Based on current practices, this implies almost a doubling of water use by agriculture worldwide.

---

* E-mail: rfleiner@icimod.org

© CAB International 2013. *Managing Water and Agroecosystems for Food Security*
(ed. E. Boelee)

However, the environmental sustainability of water use and, in many places, the limits to its absolute availability, have already been reached globally – while locally, they have even been surpassed (Molden, 2007). Many important river basins no longer have enough water for all of the human users of the resource, let alone for the environmental needs of the resource base itself, and one third of the world's population lives with physical or economic water scarcity (Molden, 2007). Furthermore, water limits have already been stretched to breaking point in important food-producing regions. For example, groundwater levels are declining rapidly in several major breadbasket and rice-bowl regions, such as the North China Plains, the Indian Punjab, and the Ogallala in Western USA (Giordano and Villholth, 2007; Shah, 2009).

Water scarcity, for both people and the environment, is related to accessibility: in regions of physical water scarcity, water is over-allocated, leaving little or none for uses that are currently given a lower priority, such as the environment. In economically water-scarce regions, water is available for use, but access is difficult, most often because of limited investment in water infrastructure (Mulligan *et al.*, 2011). When there is institutional water scarcity, both water and infrastructure are present, but national or local institutions and norms prevent some social groups or individuals from accessing water. In all cases, although in very different ways, lack of access to water is a threat to future food production and environmental sustainability, and this needs to be addressed using different approaches.

Water Availability for Agriculture

Water availability differs naturally according to agroecological zones: it is abundant in humid and sub-humid zones but is scarce in arid zones and drylands. Its availability for use is further determined by its accessibility and by the quantity that is being withdrawn, as well as by how efficiently it is being used. Estimates of the global freshwater supply are subject to high uncertainties owing to lack of available data, lack of data conformity and difficulties in data access. Even at the basin level, available water

resources are difficult to assess in detail as there are major gaps in sufficient high-quality, high-resolution and long-term hydrological data (Mulligan *et al.*, 2011). This situation is further complicated by climate change, which constitutes an additional challenge to the amount of water that is available for agriculture (Chapter 2).

Renewable freshwater expressed as long-term mean runoff has been estimated at 33,500–47,000 km^3/year (Hassan *et al.*, 2005). Fresh groundwater supply, including renewable and fossil groundwater, has been estimated to range between 7 and 23 million km^3, according to the Millennium Ecosystem Assessment (Hassan *et al.*, 2005); a more recent estimate is between 8 and 10 million km^3 (WWAP, 2012). Other water 'resources' that need to be considered are water storage and the associated infrastructure, such as dams, and natural storage such as soils, lakes, snow and ice (McCartney and Smakhtin, 2010). Water resources are unevenly distributed across the globe, and availability also depends to a large extent on the specific local context, such as physiography, land use, accessibility, infrastructure, governance, institutions and investment.

After glaciers, permanent ice and aquifers, soils are the largest store of freshwater (Hassan *et al.*, 2005) and therefore can substantially contribute to food production. However, the use efficiency of soil water depends heavily on many interacting ecophysiological processes, which ultimately determine plant growth. These processes may also be threatened by land degradation and erosion (Chapter 4).

A recent analysis of selected river basins in developing countries in different regions (those of the Andes and São Francisco in South America; of the Volta, Niger, Limpopo and Nile in Africa; and of the Karkheh, Indus-Ganges, Mekong and Yellow in Asia) showed that among these, the Mekong Basin has the relatively highest and the Limpopo and Yellow basins have the relatively lowest water balance (Mulligan *et al.*, 2011). The Yellow River is among several of the world's largest and most important rivers for socio-economic development that have been so heavily depleted and over-abstracted as to show a total lack of flow at their mouths and damage to their condition;

other examples are the Colorado and Murray–Darling Rivers (WWAP, 2009).

Infrastructure such as dams can make important contributions to development in terms of hydropower, flood control and water supply, particularly for irrigation. Notwithstanding, irrigation development has all too often come at a high environmental price tag (Faures *et al.*, 2007) and has caused the degradation of aquatic ecosystems, fragmentation and desiccation of rivers, drying up of wetlands and increased transmission of water-related diseases. Dams in particular may also affect fisheries and ecosystems in downstream areas; worldwide, the Millennium Ecosystem Assessment (Hassan *et al.*, 2005) has described around 45,000 dams larger than 15 m in height and with more than 3 million m^3 reservoir volume, and 800,000 smaller dams, with many more planned or under construction. For example, the Indus–Ganges Basin, which has an area of 0.81 Mkm^2, has 785 large dams; the Mekong Basin, which has an area of 0.54 Mkm^2, has 344 large dams; and the Yellow River Basin has 125 dams on an area of 0.86 Mkm^2 (Mulligan *et al.*, 2011). The volume of water stored in dams is estimated at 6000–7000 km^3 (Hassan *et al.*, 2005). The construction of dams and other structures along rivers has affected flows in 60% of the world's largest river systems.

The alteration of landscapes and waterscapes to increase food production has resulted in adverse, sometimes irreversible, ecological changes (Millennium Ecosystem Assessment, 2005b). For instance, intensive hydraulic infrastructure development, much of it for food production, is one of the reasons for the 35% decline in freshwater biodiversity reported between 1970 and 2005 (Hails *et al*, 2008). Reservoirs and water diversions have resulted in declining water flows and decreased sediment flows, thus preventing about 30% of sediments from reaching the sea and 10% reaching the estuaries; these sediment flows are a source of nutrients that are important for the maintenance of estuaries (Millennium Ecosystem Assessment, 2005a). Another engineering option for securing water supplies includes inter-basin water transfers; these can result in both societal costs and benefits, as well as affecting ecosystem services and biodiversity.

The extent to which water can be used for different purposes is determined by availability and access, as discussed above, but also by its quality. Despite major gaps in data and monitoring, there are indications that worldwide water quality is declining. Even though the pollution of surface waters by pathogens and organic compounds has decreased over the past 20 years in most industrial countries, water availability is threatened by water pollution in many places, for example in urban areas such as Mexico City, Delhi and Jakarta, but also in China (Millennium Ecosystem Assessment, 2005a). Nitrate concentrations have increased rapidly over the past 30 years, making this the most common chemical contaminating groundwater resources worldwide. This pollution mainly originates from pesticides, of which the USA is the largest consumer, followed by (particularly western) European countries, while Japan is the most intensive user. Excessive nitrogen application contributes to the eutrophication of freshwater and the acidification of freshwater and terrestrial ecosystems, and these can have an impact on ecosystem health and biodiversity. The capacity of ecosystems to purify such pollution is limited and the continued loss of wetlands further decreases the ability of ecosystems to filter and decompose waste (Millennium Ecosystem Assessment, 2005a; WWAP, 2012; Chapter 7).

Understanding Water Use in Agriculture

About 80% of agricultural evapotranspiration originates from rain and approximately 20% from irrigation (Molden, 2007). Estimates of total annual global freshwater withdrawals for various uses amount to 3800 km^3, of which 70% goes into food production or irrigation. Globally, water use in agriculture is larger than that for other uses, but is of relatively low value, low efficiency and highly subsidized (GWP, 2012). Furthermore, there are significant variations between countries (Molden, 2007): agricultural water use tends to decrease with increasing levels of development (WWAP, 2012). In OECD (Organisation for Economic Co-operation and Development) countries, agricultural water withdrawal accounts for 44%

of total water withdrawal, but within the eight countries that rely on irrigation it is more than 60%. For the BRIC countries (Brazil, Russian Federation, India and China), agriculture accounts for 74% of water withdrawals and ranges from 20% in the Russian Federation to 87% in India. Some fast-growing economies use up to 90% of their total freshwater withdrawal for agriculture, and least developed countries use more than 90% (WWAP, 2012). In South Asia, total renewable freshwater resources amount to 3655 km³ and total withdrawal for agriculture is 842 km³/year, which is by far the highest use of water (Atapattu and Kodituwakku, 2009). The proportion of the total actual evapotranspiration used in agriculture also varies: for example, it is as high as 67% in the Ganges Basin, 50% in the Yellow River Basin and 38% in the Mekong Basin; in contrast, it is only 6–7% in the Andes and Nile basins (Mulligan et al., 2011).

Agricultural yields range from 1.5 t/ha, for example in developing countries, to above 5–6 t/ha, as in commercial rainfed agriculture in tropical areas (Wani et al., 2009), up to around 10 t/ha. Irrigation is a well-established method of improving yield in many parts of the world and accounts for more than 40% of the increase in global food production over the past 50 years (FAO, 2011). However, in sub-Saharan Africa, for example, the use of irrigation is still low, and rainfed agriculture remains the dominant practice in subsistence agriculture. The majority of agriculture – 95% in sub-Saharan Africa and 60% in South Asia – is rainfed, but yields rarely reach 40% of their potential (Molden, 2007). Productivity from rainfed agriculture remains low as a result of limited soil nutrient availability, the occurrence of pests and diseases, and spells of minimal or no precipitation during critical growing periods. Several of these factors are related to degradation of ecosystems, and it is widely recognized that there is great potential for improvements in rainfed agriculture, which, if managed properly, could increase agricultural yields, thereby contributing to food security without additional water abstraction (see Chapter 8).

Over the past 50 years, groundwater abstraction has at least tripled, and it continues to increase by 1–2%/year, accounting for around 26% of total global water withdrawal (WWAP, 2012). Groundwater use for irrigation, estimated at 670 km³/year (as of 2010), or around two thirds of the total groundwater abstraction, is increasing, with almost 40% of irrigated areas relying on groundwater. In some countries, e.g. Saudi Arabia, nearly all of irrigation is from groundwater only (Hassan et al., 2005; FAO, 2011), and in many areas with high population density, groundwater is crucial to sustaining irrigation (Giordano and Villholth, 2007). Over-abstraction of groundwater can lead to rapid lowering of groundwater tables, such as in Yemen and in important agricultural areas of South Asia and North China, where groundwater tables have been reported to have declined at over 1 m/year (GWP, 2012).

Despite awareness of the increasing demand for food, and hence for water for agriculture, there is substantial loss from the water that is withdrawn for agriculture, via both evaporation and unsustainable use. An estimated 1210 km³/year of water is lost from groundwater and surface water sources through net evaporation from irrigation, cooling towers or reservoirs. Water loss from irrigation has been reported to account for one third of global water use (Hassan et al., 2005). Around 5–25% of global freshwater use exceeds long-term accessible supplies and is currently met either through engineered water transfers or over-abstraction of groundwater supplies; about 15–35% of irrigation withdrawals are considered to exceed supply rates and are thus unsustainable (Millennium Ecosystem Assessment, 2005a).

Livestock production systems are often considered responsible for depleting, degrading and contaminating large amounts of water (Goodland and Pimentel, 2000; Steinfeld et al., 2006). Although this view is relevant in the case of intensive and industrialized cattle systems, smallholder livestock systems have different environmental impacts (Herrero et al., 2009; Peden et al., 2009). Almost a third of global water used in agriculture is used for livestock; less than 10% of this is for drinking water, and more than 90% of it is used for feed production (Peden et al. 2007). The water in fodder that is consumed by livestock in arid and semi-arid rangelands is not readily available for other forms of agricultural production. This is especially true for ruminants such as cows and sheep (Bindraban *et al.*, 2010). With the projected increase in demand for products from livestock, agricultural water use may need to double to cater for the increased need for feed production (Chapter 2). In developing countries, the relationship between water and energy in meat that is shown in Fig. 2.4 (Chapter 2) may differ because the production of foods from animal sources can relatively easily be doubled without use of additional water by increasing livestock water productivity (Chapter 8).

Global production from inland fisheries has increased over the past 40 years, particularly from production in Asia and Africa, but in other regions – such as Europe and North America – fish production has declined as a result of environmental changes, while at the same time recreational fisheries have become more important. Productivity in aquaculture and inland capture fisheries depends on healthy ecosystems and adequate flow, quantity and quality of water (Chapter 8). Increasing human influence on freshwater bodies, e.g. through water abstraction, the building of dams, catchment management and pollution has severe impacts on the highly vulnerable fish habitat (UNEP, 2010). Freshwater fisheries are particularly threatened by water extraction or increasing water demand for other uses, and by degradation in water quality; in contrast, coastal fisheries and aquaculture are affected by increasingly nutrient-rich terrestrial runoff (Foresight, 2011). Climate change constitutes an additional risk to fish production; other drivers are changed demand, access to resources and risk margins (Bunting, 2013; Chapter 2).

Health Issues in Water and Agroecosystems

Worldwide, diseases associated with agriculture have important health impacts, particularly on poor people and those who are directly exposed to the risks, such as farmers, consumers and households in agricultural areas. Many of these health risks are related to agricultural water use (Kay, 1999; Parent *et al.*, 2002). Intensification through irrigation for productivity gains is often accompanied by increased, typically diffused, agrochemical inputs. These chemicals can also pollute waterways and pose a threat to human, livestock and ecosystem health. Further risks can derive from the toxic algal blooms that are associated with agrochemical water pollution (Chorus and Bartram, 1999).

Among the most important diseases related to agriculture are those transmitted by vectors that breed in or are associated with water, and which tend to increase as a result of irrigation and the building of dams for agricultural purposes. The most important vector-borne water-associated disease is malaria, which resulted in an estimated 655,000 deaths in 2010, with about 90% of these in Africa (Keiser *et al.*, 2005a; WHO, 2011). Other examples are lymphatic filariasis (also known as elephantiasis; Erlanger *et al.*, 2005), schistosomiasis (Steinmann *et al.*, 2006) and buruli ulcer (WHO, 2007). Many vector-borne diseases are zoonoses (animal diseases that are transmitted to people, such as sleeping sickness and Rift Valley Fever), whose presence and prevalence are linked to livestock and wildlife. Other zoonoses are transmitted via the faecal–oral route when animal faeces contaminate water that is subsequently consumed without treatment or when contaminated foods are eaten fresh; among the most notable of these are leptospirosis, salmonellosis and cryptosporidiosis, which together make tens of millions of people sick each year. Other important waterborne diseases that are neither vector borne nor zoonotic include typhoid, cholera, giardiasis, hepatitis and enteric viruses.

Waterborne zoonoses are especially likely when poorly regulated intensive livestock keeping results in the discharge of large amounts of waste to water. People who share scarce water sources with livestock and wildlife are at high risk as water storage systems can support biocoenoses in which people, livestock and wildlife are brought into close contact. This results in a greater effective contact among animals and humans, and ultimately facilitates disease transmission between animals and humans (Woodford, 2009). A study that looked into the genetic similarity of *Escherichia coli* strains from primates and humans in Uganda found that the use of water from an open water source was associated with an increased genetic similarity between the strains of these bacteria found in primates and humans (Goldberg *et al.*, 2008).

Many water-associated diseases are fostered by poorly designed or managed irrigation or water storage systems, or by harmful agricultural practices (Boelee and Madsen, 2006; Diuk-Wasser *et al.*, 2007; Boelee *et al.*, 2013). The emergence of malaria in the Thar Desert in India (Vora, 2008) and of Rift Valley fever in West Africa are notable examples (Pepin *et al.*, 2010).

The ongoing pandemic of highly pathogenic avian influenza provides a different example of the role of irrigation and livestock in disease emergence. Numerous strains of avian influenza virus of low pathogenicity circulate into the natural reservoir and in wild birds. These strains evolve towards virulence through adaptation to domestic ducks, which, through close contact, then transmit the infection to chickens. In East Asia, this is linked to rice farming combined with free grazing duck farming in wetland areas (Artois *et al.*, 2009); ducks are not very susceptible to clinical disease but they are infectious to other domestic poultry by direct contact or environmental contamination (Sims *et al.*, 2005).

Particularly in peri-urban areas, irrigation water is often contaminated with pathogens or chemicals that may affect farmers who come into contact with the water, and that may also enter the food chain, especially when crops or livestock products are eaten raw (Drechsel *et al.*, 2010). At the same time, the use of polluted water to irrigate crops supports the livelihoods of 20–50 million farmers and feeds up to a billion consumers. Water pollutants (chemical and biological) can also impair the health of livestock and of the consumers of animal products within a complex system that includes links between waterborne and food-borne vectors.

Although biological hazards are of much greater overall human health impact, agrochemicals and heavy metals can also contaminate water, and then pose a risk to human health from acute or chronic poisoning. Livestock and fish farming also lead to the presence of antibiotic residues or antibiotic-resistant bacteria in water, with potentially large impacts on human health. Many of these health problems arise from the methods by which agricultural production systems are managed and therefore could be positively influenced by an ecologically sound approach.

Improved and innovative agricultural and water management practices can help to reduce water-associated diseases (Boelee *et al.*, 2013). This reduction has to be carefully balanced with the need to support the livelihoods of farmers and provide affordable food to poor consumers. Further along the value chain, consumers can be protected and costs to the public health sector will decrease. In relation to all of the above health issues, there is vast experience of relevant agro-ecological interventions that can help to mitigate negative health impacts if water management practices are put into place (Keiser *et al.*, 2005b; McCartney *et al.*, 2007).

Similarly, a more integrated management of agroecosystems for a wider range of ecosystem services has the potential to generate additional benefits, such as enhanced pest and disease regulation. In turn, this could reduce the need for agrochemicals and limit the exposure of farmers to harmful substances, currently a significant occupational health hazard in agriculture. People in developing countries bear more than 80% of the global burden of occupational disease and injury, and the agricultural sector is one of the most hazardous (ILO, 2000). It is estimated that 2–5 million people suffer acute poisonings related to pesticides annually, 40,000 of whom die every year (Cole, 2006). Excessive use of pesticides can also lead to resistance in medically important insects. Pesticides are used inappropriately as a result of capacity deficits, inadequate regulation and perverse incentives, as well as lack of alternatives. Other agricultural inputs, such as nitrates, disinfectants, acaricides and veterinary drugs can also have negative health impacts if incorrectly used. Increased biodiversity in agroecosystems, especially when these are managed on a landscape scale and are connected by corridors that provide habitat for natural predators (Molden *et al.*, 2007), reduces the need for agrochemicals and their associated health risks.

Environmental Flows

Competition between water users has existed for millennia, especially between water abstracted for direct human well-being and water required to sustain various water-dependent ecosystem services. People with their livestock and crops have settled near water sources or seasonally migrated to access them for thousands of years, and human alteration of the structure or functioning of coastlines, rivers, lakes and other wetlands is pervasive. Water use in agriculture affects ecosystem services, not only by reducing the amounts of water available but also, among other impacts, by polluting water, altering river flow patterns and reducing habitat connectivity (Gordon and Folke, 2000).

With growing populations and increasing water use per capita, there is often not enough water of sufficient quality to meet all needs. The most common result is that non-agricultural ecosystems do not receive adequate attention and the water needs of ecosystems, or environmental flows, are not met. Consequently, important ecosystem services are often disrupted, including those related to food production, but also in the provision of clean water, fish stocks, flood control and many other functions.

Water flows dedicated to the environment, often aquatic ecosystems such as downstream rivers, have been defined as environmental flows. The most recent, widely adopted definition of an environmental flow (also referred to as the environmental or ecological water requirement, EWR) is the 'quantity, timing and quality of water flows required to sustain freshwater and estuarine ecosystems and the human livelihoods and well-being that depend on these ecosystems' (Brisbane Declaration, 2007). This definition highlights the relationship between water for ecosystem health (or 'water for nature' in integrated water resource management – IWRM) and water to sustain the livelihood needs of people, including their water and food security (there is more on the role of environmental flows in IWRM in Chapter 10).

Nowadays, agricultural water requirements can be calculated with a fair degree of accuracy, and considerable advances have similarly been made in the quantification of the water requirements of ecosystems (environmental flows) (Postel and Richter, 2003; Tharme, 2003; Poff *et al.*, 2010). This leads to reasonably accurate assessments of discrepancies between requirements and supply. In 37% of 227 large river basins that were assessed globally, environmental flows were strongly – and in 23% of them moderately – affected by fragmentation and altered flows (Cook *et al.*, 2011).

Physical water scarcity and the associated proliferation of water infrastructure are the primary causes of decreasing and altered patterns and timing of flows to ecosystems (Rosenberg *et al.*, 2000). With such impacts, ecosystems may not be able to deliver the full range of ecosystem services that are beneficial to people. Reduction in provisioning capacities can lead to economic water scarcity, which can further result in physical water scarcity once low-cost water resources are over-exploited (WWAP, 2009). Importantly for the future, a spatial analysis has shown that, at the global level, threats to water security are highly associated with threats to river-based biodiversity (Vörösmarty *et al.*, 2010). Various studies confirm that the imbalances between irrigated agriculture and nature conservation have reached a critical point on a global scale (Lemly *et al.*, 2000; Baron *et al.*, 2002). When water use for increasing agricultural production, be it crops, livestock, fisheries or aquaculture, or some combination of these, is examined in a trade-off with environmental flows, the overall food productivity from a given water resource may decrease (WWAP, 2009).

Tensions between water for ecosystems and water for food do not only have an impact on food production. Balancing the water demand for different uses is critically important for maintaining biodiversity and ecosystem resilience (WWAP, 2009), as well as for other ecosystem services – including the provision of firewood, timber, pollination services and clean water, all of which are essential for human well-being (Carpenter *et al.*, 2009). Restoring the productive capacity of highly degraded ecosystems requires revegetation and flow restoration, both of which, in turn, need water; these are needs that will compete directly with water demand for food production, as well as for other uses. In some cases, the values generated by irrigation have proved to be less than the values generated by the ecosystems they replaced (Barbier and Thompson, 1998; Acreman, 2000). In order to avoid further degradation, fundamental efforts are required to promote and establish effective ecosystem-based catchment management approaches (IUCN, 2000; UNEP, 2010; Chapter 10).

Water Availability, Poverty and Development

Water contributes to poverty alleviation in various ways, such as improving water supply and sanitation, enhancing health and resilience to disease, improving productivity and output, and helping to provide more affordable food (WWAP, 2009; Chapter 2). Generally, the poorest populations in the world face the most difficulties in accessing water supplies, and they are also the most dependent on water resources for their daily livelihoods (WWAP, 2009). Poor people also tend to be the ones that are most directly dependent on food delivery from healthy and well-functioning natural ecosystems to support often subsistence-driven livelihoods, e.g. river–floodplain fisheries and flood recession agri-culture (Richter et al., 2010). This places them as the people most vulnerable to changing conditions such as climate change, environ-mental degradation and population pressures (Molden, 2007; Sullivan and Huntingford, 2009).

While economic poverty decreased from 28% in 1990 to 19% in 2002 (UNEP, 2007), water poverty increased over that same period (WWAP, 2009). Water poverty refers to a situation where a nation or region cannot afford the cost of providing sustainable clean water to all of its people at all times (Feitelson and Chenoweth, 2002; Molle and Mollinga, 2003). This suggests that unless water poverty can be alleviated, economic poverty reduction programmes will be less effective than they would otherwise be. Increased water provision for agricultural production is thus viewed as an opportunity that allows more people to obtain an income from farming and increases food

production – thereby decreasing the overall price of food, and allows the poor to consume a more nutritional diet and spend their income on other necessities (Hussain and Hanjra, 2004; McIntyre et al., 2008).

Challenges related to water availability, water use for agriculture, and poverty are diverse and complex and therefore need to be analysed by integrating different perspectives, i.e. hydrological, water and land productivity, livelihood and development, and governance and institutions. This will allow the identification of viable options for improving the situation effectively in an integrated approach across relevant sectors. Cook et al. (2011) suggested that water is linked to development through: (i) physical water scarcity; (ii) lack of access to water, or economic water scarcity due to infrastructure or institutional frameworks, which has strong linkages to the level of development; (iii) exposure to water-related hazards such as drought, flood and disease, which are expected to be aggravated by climate change and largely affect the poor; and (iv) water productivity, which, particularly in rainfed agriculture, is low, but through improvement offers opportunities to meet future increase in demand for food without increasing agricultural water use. These relationships were analysed by Cook et al. (2011) and Kemp-Benedict et al. (2011) in a study of ten selected basins in developing countries, in which they classified the basins based upon the agricultural and water-related parameters that characterized their different levels of development (see Box 5.1).

The Yellow River Basin is a characteristic example of an area that is facing physical water scarcity; pressures on existing water resources are also expected to increase further (Ringler et al., 2010). This basin, a key food production centre of global importance, is facing growing water-related challenges, which are exacerbated by increasing demands from industrial and urban sectors, environmental needs, increasing water pollution, and potentially severe future climate change impacts that result in increasing water deficit. The main water user in the basin is agriculture, and this is also a main contributor to water pollution. Institutional and policy frameworks for managing water resources in the basin, however, are not sufficiently harmon-ized and integrated to ensure a basin-wide

---

Box 5.1. Agriculture and water in a development context (Cook *et al.*, 2011; Kemp-Benedict *et al.*, 2011).

The basins of the Andes and São Francisco (South America), Volta, Niger, Limpopo and Nile (Africa), Karkheh (Iran), Indus–Ganges (India), Mekong (South-east Asia) and Yellow (China) rivers were classified according to their water availability, water productivity and poverty. Arranged along a 'development trajectory', based on the variables of rural poverty and agriculture as a percentage of gross domestic product (GDP), the basins ranged from strongly agricultural economies, via transitional economies, to industrial economies. In this approach, agriculture is seen as a necessary but insufficient basis for development.

Agricultural economies, e.g. those of the Niger, Volta and Nile River basins are characterized by overall very low agricultural productivity and high rural poverty, limited non-agricultural economic activities and poorly developed water infrastructure; 'non-engineered' agriculture is seen as relatively more important than in the two other economies. To move along the development pathway, agricultural productivity needs to be improved, basic needs provided, markets and necessary infrastructure developed, and food security enhanced, mostly through improving rainfed agriculture rather than through irrigation.

In *transitional economies*, e.g. those of the Indus, Ganges, Mekong and Yellow River basins, non-agricultural and value-adding activities increasingly contribute to GDP and attract people from the agricultural sector. Generally, rapid economic and population growth are experienced, coupled with increasing demand for food; water resources are well developed, non-agricultural activities are expanding, but development is overall uneven and localized. Agriculture remains important at the national level, and agricultural productivity is increasing owing to the market and food demands of an increasingly urban population. As a result, pressure on water resources may increase as a result of increased agricultural activity (which possibly competes with other uses), water quality issues emerge and protection against water-related hazards is insufficient. The main opportunity for these economies is institutional development to enable transparent, informed and broadly-based processes of change for enhancing capacity and benefit sharing.

Industrial economies, e.g. those of the Andes and São Francisco River basins, no longer depend on agriculture for economic development, even if agriculture remains important in localized areas of rural poverty. Markets are well developed and income security gains more importance. In this context, water resources may be intensively managed. However, ecosystem services and benefit sharing are increasingly recognized, protection from water-related hazards is enhanced and food security is assured. There are increasing opportunities for sustaining the ecosystem services that are required to maintain water supplies to urban and industrial sectors, hydropower and agriculture.

---

approach to the sustainable management of water resources. Although measures have been taken to improve water legislation, the implementation of laws and regulations is still at a low level. Responsibilities for managing water resources in the basins are fragmented among different administrative units and levels, and these lack collaboration. Various options have been identified for addressing water scarcity in the Yellow River Basin. These include: improving water use efficiency and increasing water productivity in agriculture, industry and the domestic sector; implementing institutional solutions such as the reform of irrigation management; and using economic incentives for water management, such as pricing, taxes, subsidies, quotas and use or ownership rights. The potential for expanding irrigation in the basin is so limited that creating off-farm employment opportunities is viewed as a more appropriate strategy for advancing future rural economic development. Overall, a more consistent and harmonized approach to water resources management at the basin level, well supported by relevant institutions and policies, seems necessary to resolve the basic issues and make use of the potential of the different options that have been identified (Ringler *et al.*, 2010).

Future Challenges

The increase in water demand for food production is difficult to predict, as it depends to a large extent on variables such as population size, urbanization, diet composition, the ability to increase water use efficiency, and the effective allocation of production through enhanced trade and other means. It has been estimated that demand for water in agriculture could increase by over 30% by 2030, while total global water demand could increase by 35–60% between 2000 and 2025, and could double by 2050 owing to pressures from industry, domestic use and the need for environmental flows (Foresight, 2011). An optimized scenario accounting for regional opportunities and constraints would require global water consumption of agricultural crops to increase by 20% – or rise up to 8515 km^3 by 2050; the estimates vary depending on trade, water use efficiency, area expansion and productivity in rainfed and irrigated agriculture (de Fraiture and Wichelns, 2010). Other calculations arrive at different numbers. The 2012 World Water Development Report (WWAP, 2012) estimates an increase of around 19% by 2050, or more in the absence of technological progress or policy interventions; whereas Rockström et al. (2007) estimate that the achievement of food security in 92 developing countries would require 9660 km^3 water for agriculture by 2050. In many countries, water availability for agriculture is already limited and uncertain, and is expected to decrease further. In some arid regions, for example in the Punjab, Egypt, Libya and Australia, major non-renewable fossil aquifers are increasingly being depleted.

Climate change is expected to increasingly affect water, and hence food, security (Chapter 2), particularly in areas of Africa and Asia where agriculture depends on rainwater for its crops. In rainfed areas in Asia and sub-Saharan Africa water storage infrastructure is least developed and nearly 500 million people are at risk of food shortages (WWAP, 2012). Increased occurrence of climatic extremes such as floods and droughts can also have an impact on agricultural outputs, putting food security and economies at risk.

The expansion of irrigation across much of Asia, North America and North Africa has fuelled productivity gains in the past, but the limits have been reached, as little or no additional water is available for use in these areas (Faures et al., 2007). In these physically

water-scarce areas, but also in other regions, there will be increasing demand from other users and sectors, such as cities, industries, energy and environment, which will need to be addressed through adequate governance and institutional mechanisms.

Conclusions

Given the increasing demand and competition for freshwater resources, whose unhalted abstraction often undermines environmental flows and effects, the augmentation of water availability for agriculture can no longer rely on increasing water withdrawals, which are likely to result in further river basin closures (Molle et al., 2010) and aggravated ecosystem degradation. Instead, new approaches to water use in agriculture will need to be explored that minimize water withdrawal and its effects on the environment, such as optimized water storage in rainfed agriculture, overall increased water use efficiency, and the treatment and reuse of wastewater where possible. The greatest hope for meeting the food and water demands of the world 50 years from now probably lies in increasing agricultural water use productivity for many of the least productive areas (Molden, 2007; Chapter 8).

Sustainably meeting the agricultural water needs of a growing population will require rethinking the approach to how water is developed and managed. The challenge of the increasing water scarcity for food production and other uses must be addressed through an integrated cross-sectoral approach to water resources and ecosystems management that is linked to ecoagricultural research, and aimed at maximizing water use efficiency and productivity, and minimizing environmental and climate change impacts. Such an approach helps to sustain critical ecosystems and eco-system services, thereby supporting agricultural production and offering other multiple benefits for ecosystems, food security and human well-being.

Innovative strategies and practices will, however, need to be identified towards sustainable and integrated water resources management for various uses going beyond agriculture. The increasing imbalance between water availability for use and water withdrawn for agriculture, and also between water used in agriculture and water available for other uses, will need to be addressed through different technological, economic and institutional measures. Water conservation is a key measure that can be achieved through improving water use efficiency and water productivity in agricultural production, as well as by limiting the further expansion of water withdrawal. Sustainable pathways to growing enough food with limited water include: increasing water productivity in agriculture and, particularly, in rainfed areas; reducing loss of water in agricultural production through improved management practices and infrastructure; increasing water storage; expanding reuse of wastewater; influencing food consumption patterns; and enabling trade between water-rich and water-scarce areas. Solutions will need to be location and context specific, and adapted to the physical, economic and sociocultural environment. Countries must consider the full social, economic and environmental costs of not conserving existing water resources, as well as the costs of failure to develop new water sources.

Facilitating sustainable and more effective water resources management depends to a large extent on the governance and institutional frameworks that are in place. To enhance equal development and the sharing of resources and benefits, institutions are required to develop an integrated approach across different relevant sectors; this will enable them to balance the demands of different groups of people, as well as the pressures for development and sustainable use of the natural environment (Cook *et al.*, 2011), and also to put the necessary regulatory and incentive mechanisms in place. In the years to come, improvements in the collection of the data relevant to water availability, and to climate change, population growth and development, will help to provide a better basis for informed decision making.

References

Acreman, M. (2000) Background study for the World Commission of Dams. Reported in: World Commission on Dams (2000) *Dams and Development: A New Framework for Decision-making.* Earthscan Publications, London and Sterling, Virginia.

Artois, M., Bicout, D., Doctrinal, D., Fouchier, R., Gavier-Widen, D., Globig, A., Hagemeijer, W., Mundkur, T., Munster, V. and Olsen, B. (2009) Outbreaks of highly pathogenic avian influenza in Europe: the risks associated with wild birds. *Revue Scientifique et Technique – Office International des Épizooties* 28, 69–92.

Atapattu, S.S. and Kodituwakku, D.C. (2009) Agriculture in South Asia and its implications on downstream health and sustainability: a review. *Agricultural Water Management* 96, 361–373. doi:10.1016/j.agwat.2008.09.028

Barbier, E.B. and Thompson, J.R. (1998) The value of water: floodplain versus large-scale irrigation benefits in northern Nigeria. *Ambio* 27, 434–440.

Baron, J.S., Poff, L.N., Angermeier, P.L., Dahm, C.N., Gleick, P.H., Hairston, N.G. Jr, Jackson, R.B., Johnston, C.A., Richter, B.D. and Steinman, A.D. (2002) Meeting ecological and societal needs for freshwater. *Ecological Applications* 12, 1247–1260. doi:10.1890/1051-0761(2002)012[1247:MEASNF]2.0.CO;2

Bindraban, P., Conijn, S., Jongschaap, R., Qi, J., Hanjra, M., Kijne, J., Steduto, P., Udo, H., Oweis, T. and de Boer, I. (eds) (2010) *Enhancing Use of Rainwater for Meat Production on Grasslands: An Ecological Opportunity Towards Food Security. Proceedings 686, International Fertiliser Society, Cambridge, 10 December 2010, Leek, UK.* ISRIC World Soil Information, Wageningen, the Netherlands.

Boelee, E. and Madsen, H. (2006) *Irrigation and Schistosomiasis in Africa: Ecological Aspects.* IWMI Research Report 99, International Water Management Institute, Colombo. doi:10.3910/2009.099

Boelee, E., Yohannes, M., Poda, J.-N., McCartney, M., Cecchi, P., Kibret, S., Hagos, F. and Laamrani, H. (2013) Options for water storage and rainwater harvesting to improve health and resilience against climate change in Africa. *Regional Environmental Change* 13(3), 509–519. doi.org/10.1007/s10113-012-0287-4

Brisbane Declaration (2007) *The Brisbane Declaration: Environmental Flows are Essential for Freshwater Ecosystem Health and Human Well-Being.* Declaration made at: 10th International River Symposium and Environmental Flows Conference, Brisbane, Australia, 3–6 September 2007. Available at: http://www.eflownet.org/download_documents/brisbane-declaration-english.pdf (accessed February 2013).

Bunting, S.W. (2013) *Principles of Sustainable Aquaculture: Promoting Social, Economic and Environmental Resilience.* Earthscan, London, from Routledge, Oxford, UK.

Butchart, S.H.M. *et al.* (2010) Global biodiversity: indicators of recent declines. *Science* 328, 1164–1168. doi:10.1126/science.1187512

Cai, X. and Rosegrant, M.W. (2002) Global water demand and supply projections. Part 1. A modelling approach. *Water International* 27, 159–169.

Carpenter, S.R., Mooney, H.A., Agard, J., Capistrano, D., DeFries, R.S., Díaz, S., Dietz, T., Duraiappah, A.K., Oteng-Yeboah, A., Pereira, H.M., Perrings, C., Reid, W.V., Sarukhan, J., Scholes, R.J. and Whyte, A. (2009) Science for managing ecosystem services: beyond the Millennium Ecosystem Assessment. *Proceedings of the National Academy of Sciences of the United States of America* 106, 1305–1312. doi:10.1073/pnas.0808772106

Chorus, I. and Bartram, J. (eds) (1999) *Toxic Cyanobacteria in Water. A Guide to their Public Health Consequences, Monitoring, and Management.* E. and F.N. Spon (Routledge), London/New York for World Health Organization, Geneva, Switzerland.

Cole, D. (2006) Occupational health hazards of agriculture. In: Hawkes, C. and Ruel, M.T. (eds) *Understanding the Links Between Agriculture and Health. 2020 Vision Focus 13 for Food, Agriculture and the Environment.* International Food Policy Research Institute, Washington, DC, Brief 8 of 16, May 2006, pp. 17–18.

Cook, S., Fisher, M., Tiemann, T. and Vidal, A. (2011) Water, food and poverty: global- and basin-scale analysis. *Water International* 36, 1–16. doi:10.1080/02508060.2011.541018

de Fraiture, C. and Wichelns, D. (2010) Satisfying future demands for agriculture. *Agricultural Water Management* 97, 502–511. doi:10.1016/j.agwat.2009.08.008

Diuk-Wasser, M.A., Touré, M.B., Dolo, G., Bagayoko, M., Sogoba, N., Sissoko, I., Traoré, S.F. and Taylor, C.E. (2007) Effect of rice cultivation patterns on malaria vector abundance in rice-growing villages in Mali. *American Journal of Tropical Medicine and Hygiene* 76, 869–874.

Drechsel, P., Scott, C.A., Raschid-Sally, L., Redwood, M. and Bahri, A. (eds) (2010) *Wastewater Irrigation and Health: Assessing and Mitigating Risk in Low-income Countries.* Earthscan, London, International Water Management Institute (IWMI), Colombo and International Development Research Centre (IDRC), Ottawa, Canada.

Erlanger, T.E., Keiser, J., Caldas De Castro, M., Bos, R., Singer, B.H., Tanner, M. and Utzinger, J. (2005) Effect of water resource development and management on lymphatic filariasis, and estimates of populations at risk. *American Journal of Tropical Medicine and Hygiene* 73, 523–533.

FAO (2011) *The State of the World's Land and Water Resources for Food and Agriculture (SOLAW) – Managing Systems at Risk (Summary Report).* Food and Agriculture Organization of the United Nations, Rome and Earthscan, London.

Faures, J.M., Svendsen, M. and Turral, H. *et al.* (2007) Reinventing irrigation. In: Molden, D. (2007) *Water for Food, Water for Life: Comprehensive Assessment of Water Management in Agriculture.* Earthscan, London, in association with International Water Management Institute (IWMI), Colombo, pp. 353–394.

Feitelson, E. and Chenoweth, J. (2002) Water poverty: towards a meaningful indicator. *Water Policy* 4, 263–281. doi:10.1016/S1366-7017(02)00029-6

Foresight (2011) *The Future of Food and Farming: Challenges and Choices for Global Sustainability. Final Project Report.* The Government Office for Science, London.

Giordano, M. and Villholth, K. (eds) (2007) *The Agricultural Groundwater Revolution: Opportunities and Threats to Development.* Comprehensive Assessment of Water Management in Agriculture Series 3, CAB International, Wallingford, UK.

Goldberg, T.L., Gillespie, T.R., Rwegio, I.B., Estoff, E.L. and Chapman, C.A. (2008) Forest fragmentation as cause of bacterial transmission among nonhuman primates, humans, and livestock, Uganda. *Emerging Infectious Diseases* 14, 1375–1382.

Goodland, R. and Pimentel, D. (2000) Environmental sustainability and integrity in the agriculture sector. In: Pimentel, D., Westra, L. and Noss, R.F. (eds) *Ecological Integrity: Integrating Environment, Conservation and Health.* Island Press, Washington, DC, pp. 121–38.

Gordon, L. and Folke, C. (2000) Ecohydrological landscape management for human well-being. *Water International* 25, 178–184. doi:10.1080/02508060008686816

GWP (2012) *Increasing Water Security – A Development Imperative.* Perspectives Paper. Global Water Partnership, Stockholm. Available at: http://www.gwp.org/Global/About%20GWP/Publications/Perspectives%20Paper_Water%20Security_final.pdf (accessed February 2013).

Hails, C., Humphrey, S., Loh, J. and Goldfinger, S. (eds) (2008) *Living Planet Report 2008.* World Wide Fund For Nature (WWF), Gland, Switzerland.

Hassan, R., Scholes, R. and Ash, N. (eds) (2005) *Ecosystems and Human Well-being: Current State and Trends, Volume 1. Findings of the Condition and Trends Working Group of the Millennium Ecosystem Assessment.* World Resources Institute and Island Press, Washington, DC.

Herrero, M., Thornton, P.K., Gerber, P. and Reid, R.S. (2009) Livestock, livelihoods and the environment: understanding the trade-offs. *Current Opinion in Environmental Sustainability* 1, 111–120. doi:10.1016/j.cosust.2009.10.003

Hussain, I. and Hanjra, M. (2004) Irrigation and poverty alleviation: review of the empirical evidence. *Irrigation and Drainage* 53, 1–15.

ILO (2000) *Safety and Health in Agriculture. International Labor Conference, 88th Session, May 30–June 15.* International Labour Organization, Geneva, Switzerland.

IUCN (2000) *Vision for Water and Nature: A World Strategy for Conservation and Sustainable Management of Water Resources in the 21st Century.* IUCN – The World Conservation Union, Gland, Switzerland.

Kay, B.H. (ed.) (1999) *Water Resources: Health, Environment and Development.* E. and F.N. Spon (Routledge), London/New York.

Keiser, J., Caldas de Castro, M., Maltese, M.F., Bos, R., Tanner, M., Singer, B.H. and Utzinger, J. (2005a) Effect of irrigation and large dams on the burden of malaria on a global and regional scale. *American Journal of Tropical Medicine and Hygiene* 72, 392–406.

Keiser, J., Singer, B.H. and Utzinger, J. (2005b) Reducing the burden of malaria in different eco-epidemiological settings with environmental management: a systematic review. *The Lancet Infectious Diseases* 5, 695–708.

Kemp-Benedict, E., Cook, S., Allen, S.L., Vosti, S., Lemoalle, J., Giordano, M., Ward, J. and Kaczan, D. (2011) Connections between poverty, water and agriculture: evidence from 10 river basins. *Water International* 36, 125–140. doi:10.1080/02508060.2011.541015

Lemly, A.D., Kingsford, R.T. and Thompson, J.R. (2000) Irrigated agriculture and wildlife conservation: conflict on a global scale. *Environmental Management* 25, 485–512. doi:10.1007/s002679910039

McCartney, M. and Smakhtin, V. (2010) *Water Storage in an Era of Climate Change: Addressing the Challenge of Increasing Rainfall Variability.* IWMI Blue Paper, International Water Management Institute, Colombo.

McCartney, M.P., Boelee, E., Cofie, O. and Mutero, C.M. (2007) *Minimizing the Negative Environmental and Health Impacts of Agricultural Water Resources Development in Sub-Saharan Africa.* IWMI Working Paper 117, International Water Management Institute, Colombo. doi:10.3910/2009.297

McIntyre, B.D., Herren, H.R., Wakhungu, J. and Watson, R.T. (eds) (2008) *Agriculture at a Crossroads. International Assessment of Agricultural Knowledge, Science and Technology for Development (IAASTD): Global Report.* Island Press, Washington, DC. Available at: http://www.agassessment.org/reports/IAASTD/EN/Agriculture%20at%20a%20Crossroads_Global%20Report%20(English).pdf (accessed February 2013).

Millennium Ecosystem Assessment (2005a) *Ecosystems and Human Well-being: Wetlands and Water – Synthesis. A Report of the Millennium Ecosystem Assessment.* World Resources Institute, Washington, DC. Available at: www.maweb.org/documents/document.358.aspx.pdf (accessed February 2013).

Millennium Ecosystem Assessment (2005b) *Living Beyond Our Means. Natural Assets and Human Well-being. Statement from the Board.* Available at: http://www.unep.org/maweb/documents/document.429.aspx.pdf (accessed February 2013).

Molden, D. (ed.) (2007) *Water for Food, Water for Life: Comprehensive Assessment of Water Management in Agriculture.* Earthscan, London, in association with International Water Management Institute (IWMI), Colombo.

Molden, D., Tharme, R., Abdullaev, I. and Puskur, R. (2007) Managing irrigation systems. In: Scherr, S.J. and McNeely, J.A. (eds) *Farming with Nature. The Science and Practice of Ecoagriculture.* Island Press, Washington, DC, pp. 231–249.

Molle, F. and Mollinga, P. (2003) Water poverty indicators: conceptual problems and policy issues. *Water Policy* 5, 529–544.

Molle, F., Wester, P. and Hirsch, P. (2010) River basin closure: processes, implications and responses. *Agricultural Water Management* 97, 569–577. doi:10.1016/j.agwat.2009.01.004

Mulligan, M., Saenz Cruz, L.L., Pena-Arancibia, J., Pandey, B., Mahé, G. and Fisher, M. (2011) Water availability and use across the Challenge Program on Water and Food (CPWF) basins. *Water International* 36, 17–41.

Parent, G., Zagré, N.-M., Ouédraogo, A. and Guiguembé, R.T. (2002) Les grands hydro-aménagements au Burkina Faso contribuent-ils à l'amélioration des situations nutritionelles des enfants? [Do large hydroelectric facilities in Burkina Faso contribute to improving the nutritional situation of children?] *Cahiers Agricultures* 11, 51–57.

Peden, D., Tadesse, G. and Misra, A.K. *et al.* (2007) Water and livestock for human development. In: Molden, D. (ed.) *Water for Food, Water for Life: Comprehensive Assessment of Water Management in Agriculture.* Earthscan, London, in association with International Water Management Institute (IWMI), Colombo, pp. 485–514.

Peden, D., Alemayehu, M., Amede, T., Awulachew, S.B., Faki, H., Haileslassie, A., Herero, M., Mapezda, E., Mpairwe, D., Musa, M.T., Taddesse, G. and van Breugel, P. (2009) *Nile Basin Livestock Water Productivity, Project Number 37.* CPWF Project Report, CGIAR Challenge Program on Water and Food, Colombo.

Pepin, M., Bouloy, M., Bird, B.H., Kemp, A. and Paweska, J. (2010) Rift Valley fever virus (*Bunyaviridae: Phlebovirus*): an update on pathogenesis, molecular epidemiology, vectors, diagnostics and prevention. *Veterinary Research* 41, 61.

Poff, N.L. *et al.* (2010) The ecological limits of hydrologic alteration (ELOHA): a new framework for developing regional environmental flow standards. *Freshwater Biology* 55, 147–170. doi:10.1111/j.1365-2427.2009.02204.x

Postel, S. and Richter, B. (2003) *Rivers for Life: Managing Water for People and Nature.* Island Press, Washington DC.

Richter, B.D., Postel, S., Revenga, C., Scudder, T., Lehner, B., Churchill, A. and Chow, M. (2010) Lost in development's shadow: the downstream human consequences of dams. *Water Alternatives* 3, 14–42.

Ringler, C. *et al.* (2010) Yellow River basin: living with scarcity. *Water International* 35, 681–701.

Rockström, J., Lannerstad, M. and Falkenmark, M. (2007) Assessing the water challenge of a new green revolution in developing countries. *Proceedings of the National Academy of Sciences of the United States of America* 104, 6253–6260. doi:10.1073/pnas.0605739104

Rosenberg, D.M., McCully, P. and Pringle, C.M. (2000) Global-scale environmental effects of hydrological alterations: introduction. *BioScience* 50, 746–751.

Shah, T. (2009) *Taming the Anarchy: Groundwater Governance in South Asia.* Resources for the Future, Washington, DC and International Water Management Institute (IWMI), Colombo.

Sims, L.D., Domenech, J., Benigno, C., Kahn, S., Kamata, A., Lubroth, J., Martin, V. and Roeder, P. (2005) Origin and evolution of highly pathogenic H5N1 avian influenza in Asia. *Veterinary Record* 157, 159–164.

Steinfeld, H., Gerber, P., Wassenaar, T., Castel, V., Rosales, M. and de Haan, C. (2006) *Livestock's Long Shadow: Environmental Issues and Options.* Food and Agriculture Organization of the United Nations, Rome. Available at: www.fao.org/docrep/010/a0701e/a0701e00.HTM (accessed December 2012).

Steinmann, P., Keiser, J., Bos, R., Tanner, M. and Utzinger J. (2006) Schistosomiasis and water resources development: systematic review, meta-analysis, and estimates of people at risk. *The Lancet Infectious Diseases* 6, 411–425. doi:10.1016/S1473–3099(06)70521–7

Strzepek, K. and Boehlert, B. (2010) Competition for water for the food system. *Philosophical Transactions of the Royal Society B* 365, 2927–2940. doi:10.1098/rstb.2010.0152

Sullivan, C. and Huntingford, C. (2009) Water resources, climate change and human vulnerability. In: Anderssen, R.S., Braddock, R.D. and Newman L.T.H. (eds) *18th World IMACS Congress and MODSIM09 International Congress on Modelling and Simulation. Modelling and Simulation Society of Australia and New Zealand and International Association for Mathematics and Computers in Simulation, Cairns, Australia, 13–17 July 2009,* pp. 3984–3990. Available at: http://www.mssanz.org.au/modsim09/I13/sullivan_ca.pdf (accessed February 2013).

Tharme, R.E. (2003) A global perspective on environmental flow assessment: emerging trends in the development and application of environmental flow methodologies for rivers. *River Research and Application* 19, 397–441. doi:10.1002/rra.736

UNEP (2007) *Global Environment Outlook. GEO-4. Environment for Development.* United Nations Environment Programme, Nairobi. Available at: www.unep.org/geo/geo4.asp (accessed December 2012).

UNEP (2010) *Blue Harvest: Inland Fisheries as an Ecosystem Service.* WorldFish, Penang, Malaysia and United Nations Environment Programme, Nairobi.

Vora, N. (2008) Impact of anthropogenic environmental alterations on vector-borne diseases. *Medscape Journal of Medicine* 10, 238–245.

Vörösmarty, C.J., McIntyre, P.B., Gessner M.O., Dudgeon, D., Prusevich, A., Green, P., Glidden, S., Bunn, S.E., Sullivan, C.A., Reidy Liermann, C. and Davies, P.M. (2010) Global threats to human water security and river biodiversity. *Nature* 467, 555–561. doi:10.1038/nature09440

Wani, S.P., Rockström, J. and Oweis, T. (eds) (2009) *Rainfed Agriculture: Unlocking the Potential.* Comprehensive Assessment of Water Management in Agriculture Series 7. CABI, Wallingford, UK; International Crops Research Institute for the Semi-Arid Tropics (ICRISAT), Patancheru, Andhra Pradesh, India; International Water Management Institute (IWMI) Colombo, Sri Lanka.

WHO (2007) *Buruli Ulcer Disease (Mycobacterium ulcerans Infection).* Fact Sheet 1999, World Health Organization, Geneva, Switzerland.

WHO (2011) *World Malaria Report 2011.* World Health Organization, Geneva, Switzerland. Available at: www.who.int/malaria/world_malaria_report_2011/en/ (accessed December 2012).

Wisser, D., Frolking, S., Douglas, E.M., Fekete, B.M., Vörösmarty, C.J. and Schumann, A.H. (2008) Global irrigation water demand: variability and uncertainties arising from agricultural and climate data sets. *Geophysical Research Letters* 35(24): L24408. doi:0.1029/2008GL035296

Woodford, M.H. (2009) Veterinary aspects of ecological monitoring: the natural history of emerging infectious diseases of humans, domestic animals and wildlife. *Tropical Animal Health Production* 41, 1023–1033.

WWAP (2009) *The United Nations World Water Development Report 3 (WWDR3). Water in a Changing World.* World Water Assessment Programme, United Nations Educational, Scientific and Cultural Organization (UNESCO), Paris and Earthscan, London. Available at: http://unesdoc.unesco.org/images/0018/001819/181993e.pdf#page=5 (accessed February 2013).

WWAP (2012) *The United Nations World Water Development Report 4 (WWDR4): Managing Water Under Uncertainty and Risk. Executive Summary.* United Nations Educational, Scientific and Cultural Organization (UNESCO), Paris. Available at: http://unesdoc.unesco.org/images/0021/002171/217175e.pdf (accessed February 2013).

6 Drylands

Elaine M. Solowey,[1]* Tilahun Amede,[2] Alexandra Evans,[3] Eline Boelee[4] and Prem Bindraban[5]

[1]*The Arava Institute for Environmental Studies (AIES), Hevel Eilot, Israel;* [2]*International Crops Research Institute for the Semi-arid Tropics (ICRISAT), Maputo, Mozambique;* [3]*Edge Grove School, Aldenham Village, Watford, UK;* [4]*Water Health, Hollandsche Rading, the Netherlands;* [5]*World Soil Information (ISRIC) and Plant Research International, Wageningen, the Netherlands*

Abstract

Drylands are characterized by physical water scarcity, often associated with land degradation and desertification. Other factors that contribute to these problems include high population densities, unwise agricultural practices and overgrazing. However, while desert ecosystems are fragile and vulnerable and can collapse in the short term, given the right conditions and protection, these areas also have a great potential for recovery. Examples of the recovery of areas have led to the formation of counter paradigms and the emergence of a new understanding of drylands. This new understanding is founded on the recognition of the variability of these ecosystems from place to place and year to year, and of the influences of desert plants, animals and the agricultural practices of the people who live in drylands. This chapter defines both old and new paradigms, and discusses conditions that lead to non-sustainable situations and vulnerabilities. In addition, strategies are considered that can lead to proper land use and recovery.

Background

Drylands are arid and semi-arid areas where evapotranspiration exceeds rainfall for some part of the year but where there are still opportunities for livestock raising and seasonal cropping. These lands are found on all continents and include roughly all of the Middle East, half of India and about 70% of Africa (including the millet-based Sudano–Sahelian zone, the maize–groundnut belt of southern Africa and the Maghreb). There is evidence to support the idea that the actual land mass that can be considered arid or semi-arid is growing (UNCCD, 2010). Physical water scarcity, probably the most prominent constraint in drylands, is worsening, with per capita water flows reduced by many biophysical and social factors. This physical water scarcity is tied to reduced rainfall intensity, uneven distribution of rainfall with frequent drought cycles, and poor soil water holding capacity of the

* E-mail: elaine.solowey@arava.org

landscapes. Water scarcity for agriculture is also due to poor water management and poor agricultural practices that lead to low soil moisture contents, low plant productivity, low nutrient availability and poor soil development. In turn, this results in a relatively high susceptibility to soil erosion, salinization and land degradation in general (Millennium Ecosystem Assessment, 2005; Chapter 4).

Physical water scarcity in drylands is mostly linked to climate variability and recurrent droughts, which cause variations in primary production. Climate change, together with decreasing amounts of rainfall and increasing rainfall variability (Burke *et al*., 2006), is believed to exacerbate these constraints, especially for those who do not have secure access to irrigation water. High population growth rates in drylands, especially in the tropics, has led to land use changes in the entire watershed – from the water towers (mountain areas) to the lowlands – that might trigger land degradation if supportive institutional and sociopolitical mechanisms are not present.

Challenges

Desertification, defined as resource (land, water, vegetation, biodiversity) degradation, is a major environmental problem in drylands, impairing various ecosystem services. It is related to the inherent vulnerability of the land and is caused by a combination of social, economic and biophysical factors, operating at varying scales. The direct effects of desertification include soil nutrient losses, decreased infiltration and soil water holding capacity, and impaired primary productivity. These, in turn, result in changes in the species of plants and animals that can survive in the area, as well as in the disruption of various ecosystem services, including nutrient cycling, water regulation and provision, and climate regulation (Millennium Ecosystem Assessment, 2005). Biodiversity, which is key to the provision of various dryland ecosystem services, decreases as a result of land degradation. According to the desertification paradigm, which is based on the assumption that natural systems are in a state of equilibrium that can be irreversibly disrupted (Millennium Ecosystem Assessment, 2005), desertification leads to a downwards spiral of productivity loss and increasing poverty.

However, evidence of recovery in areas that were previously thought to be irreversibly degraded, e.g. the greening of the Sahel (Herrmann *et al*., 2005; Olsson *et al*., 2005), has led to the emergence of counter-paradigms. Some argue that dryland agroecosystems are better described as non-equilibrium systems, in which considerable variability from place to place and from year to year is common, and related to irregular events, such as droughts, that impede the establishment of stable states (Ellis and Swift, 1988; Behnke *et al*., 1993). Others suggest that 'triggers' must be found in order to enable the rapid rehabilitation of degraded areas. For example, in northern Uganda, Mugerwa (2009) found a solution for overcoming the tendency for termites to keep degraded rangelands in a state of non-productivity. There is an emerging consensus that both dryland ecology (Scheffer *et al*., 2001; Washington-Allen and Salo, 2007) and people's livelihoods (Folke, 2006) in dry areas respond to key drivers of change in a non-linear way, so that systems have multiple states displaying some sort of stability, which are separated by thresholds. State and transition models (Stringham *et al*., 2003) have begun to replace models based on equilibrium concepts, and diagnostic tools for detecting thresholds using remote sensing are being developed and applied (Washington-Allen *et al*., 2008).

The main objective of sustainable agriculture in drylands is to produce crops and feed livestock in a manner that utilizes the limited water resources efficiently, without applying harmful methods of cultivation and without overgrazing or otherwise endangering fragile marginal lands. Conventional agriculture from milder climates that requires expensive inputs to produce fruits and vegetables is rarely sustainable in arid zones. In fact, conventional water-intensive agricultural methods in arid zones may deplete water resources beyond their recovery capacity, sometimes until the resources are no longer usable, and greatly contribute to soil loss by water and wind erosion. Vegetation depletion, the loss of potentially valuable species of plants, and the loss of fertility and productivity in marginal lands under cultivation are also contributing factors.

Therefore, more appropriate approaches for drylands must be applied, based on both cultivating and protecting dryland agro-ecosystems. Examples include the replanting of degraded areas with useful plants that are tolerant or resistant to drought and salinity, cultivation in soil and water-thrifty modes, or managing grazing and water collection areas with an eye to conservation and future use. The greening of the Sahel after successive droughts was attributed partly to increased rain but also to widespread adoption of sustainable farming practices, such as the laborious plant-ing of windbreaks and shelterbelts, the establishment of resilient plants and field texturing (such as making contour bunds and ditches) (Herrmann et al., 2005; Reij et al., 2005, 2009).

With increasing population pressure, traditional agriculture may no longer be sufficient to maintain the productivity of arid ecosystems. Sustainable agriculture in arid and saline areas must thus be based on an integrated approach that maximizes technical opportunities for the development of specifically desert-adapted crops, soil fertility improvement, protecting fragile desert soil, integrating local crops and animals, and mobilizing underutilized water sources. Employing rainwater manage-ment strategies at plot, farm and landscape scales is a valid entry point for rehabilitating the vegetation and improving the productivity of these dryland systems, especially if soil storage systems can also be employed. The synergy of such a combined strategy will greatly increase the use efficiency of the resource base. The expert use of local inputs, local knowledge and indigenous crops, utilized with an eye to the conservation of desert soil and the thrifty use of all appropriate water resources, can enhance local agricultural systems and increase their ability to support local people (both women and men).

Such an approach would not preclude the cultivation of livelihood crops or plough agriculture but would integrate the conventional crops into rotations and reclamation projects to allow greater sustainability (Kirkby et al., 1995). Protecting degraded landscapes from direct contact with livestock and people for a limited period of time has been found to be an effective strategy for returning landscapes to productivity in Ethiopia (Amede et al., 2011; see also Box 9.1 in Chapter 9). In addition, a broader approach to agroecosystem manage-ment increases the options for livelihoods and employment at the local level, especially for women, by creating opportunities for trade and processing, and by increasing the amount of usable materials for the dryland household.

The enhancement of existing farming systems or the introduction of new ones requires the integration of the different needs, interests and perceptions of local male and female farmers, particularly of marginal groups who are more vulnerable to environmental degradation. These management strategies also seek collective action at community and higher levels to facilitate the interaction of system components and to combine pro-duction with sustainable resources manage-ment. The successful experiences of the Globally Important Agricultural Heritage Systems (GIAHS) initiated by FAO (Food and Agriculture Organization of the United Nations) in 2002 in the drylands of Morocco, Italy and the USA demonstrate the importance of global support to local indigenous knowledge systems in preserving the productivity of these arid landscapes (GIAHS, 2013). Agricultural changes might trigger different impacts on the livelihoods of men and women, and on small and large landholders, whose diversity needs to be taken into account. New crops, new technologies and external inputs such as soil fertilization may be required to optimize the agroecosystem and produce food sustainably. Where feasible, these approaches can be fitted carefully around traditional agricultural practices to make more water available, including through the development of groundwater resources, which can lead to the synergistic integration of agriculture, animal husbandry, conservation planting and agroforestry.

Dryland Soil Management

Topsoil is a resource that is formed and renewed very slowly in drylands. Low levels of macronutrients, nitrogen, phosphorus and potassium are not a problem unless the soil surface has been lost (Bainbridge, 2007), but nutrients are often concentrated in the top 2–3 cm of desert soil. Newly cultivated dryland soils often produce a sudden and one-time flush of fertility, setting an excellent crop, but as the accumulated organic material is used up, the nutrients are depleted and the soil becomes compacted, and further yields are usually disappointing.

A specific risk in drylands is the development of impermeable clay crusts when the clay, which is normally dispersed throughout the soil profile, is dissolved by excess water and floats to the top when water pools; later, when the water evaporates, the clay hardens in the sun to hard ceramic-like plates on the soil surface. Compaction and disturbance, as happens with frequent ploughing, also reduce the populations of beneficial soil organisms. The total numbers of fungi, bacteria and nematodes tend to be much lower in disturbed soils, while pathogens are more common and the soil regenerative influences of ant and termite colonies are greatly reduced.

Without sufficient protective land cover, wind erosion can move vast quantities of soil away and up into the air, causing choking storms, burying plants and crops, and contaminating food and water (there is more on land degradation and soil erosion in Chapter 4). Entire communities can disappear in eroded dryland areas under layers of sand and dust, as happened in the infamous Dust Bowl in the USA in the 1930s, and in the serious and ongoing encroachment of the sands of the Gobi Desert on to agricultural land in China.

With adequate management, it is possible to build up and protect topsoil and so enhance the supporting and regulatory services of the ecosystem, e.g. nitrogen can be increased by the planting of nitrogen fixing trees and legumes, and by the utilization of manure. Such ecological practices can help to prevent and reduce erosion. Available potassium is increased with the breakdown of plant materials, especially leaf litter. Phosphorus in desert soils is often bound up in unusable forms, and the nutrient is released only by biological activity in the soil. The application of organic fertilizers such as manure and compost can increase the soil's water holding capacity in addition to providing plant nutrition.

The strategies that best address the problems of erosion are those that lessen the force of the wind, combined with techniques that slow and hold the water so that it can be used to stimulate vegetation. Both water and wind erosion can thus be addressed by approaches that entail a certain amount of field texturing and the planting of especially hardy types of plants and trees. Techniques like these have also been proposed as part of an ecosystem approach to land and water management in the Tana River Basin in Kenya, particularly in the drier middle catchment (Knoop *et al.*, 2012).

Soil building can be enhanced by improved nutrient cycling (see Chapter 4), particularly through improved crop–livestock linkages and reclamation plantings that encourage soil microorganisms. The rational use of combined interventions from modern and traditional desert agriculture can offer new ways to cultivate the desert in a sustainable manner.

Mobilizing Water in Drylands

In an effort to supply the needs of the populations in drylands for water, food and produce, various forms of rainwater management practices have been initiated in several countries (e.g. Ngigi, 2003; Vohland and Boubacar, 2009). For instance, traditional 'tanks' in South Asia, small water harvesting structures in West Africa (zai pits, small reservoirs), soil and water conservation practices in Ethiopia, and groundwater use in Southern Africa are examples of cases where improved water management practices are bringing about change in people's livelihoods. Runoff, wastewater (including grey and black water, treated and untreated) and saline water resources are being used for farming (see Box 6.1). Saline or brackish water, often of a quality that precludes drinking, is a commonly underutilized water resource in many areas, although it can only be used for carefully selected crops, and in agricultural strategies such as the cultivation of halophytic annuals or perennials, or local grass or green manure crops that are salt tolerant. In areas lacking

---

Box 6.1. Examples of water collection in arid areas

Runoff water can be directed after collection, via a division box, to lateral canals, especially across the face of a slope to allow for storage in that slope, or it can be directed into small depressions or ditches in more level areas (Knoop *et al.*, 2012). These features can be produced by hand labour with simple tools. Both slopes and ditches can be planted with perennials that have low water use and heavily mulched to prevent evaporation. Water can also be stored in contour bunds or grass strips by directing it into loading ditches on the upslope side of the features. A strip planted with grass or a fodder crop will wick the water laterally across its face, while an elevated planted bed formation will absorb the water upwards into its core. A sound combination of interventions could also help to protect against wind and water erosion.

 These principles have been applied in a rainwater harvesting project managed by The Arava Institute for Environmental Studies (AIES) in the Negev, Israel. Nir Moshe, with an average annual rainfall of 250 mm, is the site of AIES's largest rainwater collection experiment in the Negev. This project has 20,000 m² of contour bunds planted with drought-tolerant trees, and contour furrows that collect rainwater from a series of nearby slopes. A pond has been created at what was once the lowest point of a gully caused by erosion by closing the gully at one end, and then gravelling and lining it so that it can accommodate several thousand cubic metres of water. By the end of January 2010, after one winter in operation, 2,500,000 l of water had been collected on this site by the catchment furrows and stored in tree-covered bunds. The runoff water was drained into the small pond. This rainwater harvesting system feeds an agroecosystem that provides a range of provisioning (food, fodder and other products from the trees), regulatory (water and erosion regulation) and supporting (nutrient cycling) ecosystem services.

---

reservoir sites or ponds for natural water storage, soil-based storage of moisture is an interesting possibility and can be done by, for example, improved *in situ* water management and groundwater recharge (Johnston and McCartney, 2010; McCartney and Smakhtin, 2010).

 The use of wastewater in agriculture is a common practice in many countries, often in response to water shortages or changes in water supply and demand, or because traditional sources of irrigation water have been polluted with effluent. Estimates of wastewater use vary, not least because there is no agreed classification system, but some 23 countries use untreated wastewater, 20 use treated wastewater and a further 20 use both types of wastewater (Jiménez *et al.*, 2010). FAO estimates that wastewater is used on 10% of all irrigated land (FAO, 2009; Winpenny *et al.*, 2010). Much of the planned use of treated wastewater irrigation is currently in arid areas, for example Israel, which is a world leader in reclaiming more than 60% of its sewage effluent (Hamilton *et al.*, 2007). Furthermore, Scott *et al.* (2010) estimate that the area that uses wastewater informally is ten times larger than that which uses it formally (Drechsel *et al.*, 2011).

The drivers of wastewater irrigation are complex, but they include access to a secure, year-round source of water (as well as nutrients) that allows farmers to irrigate in the dry season and supplement their incomes. In some cases, wastewater use has arisen because the supply of traditional water resources, such as canal water in Pakistan, has diminished over the years (Weckenbrock *et al.*, 2011) or have become polluted. The result is that wastewater use is an important part of agricultural production throughout the world and it should be considered as a legitimate component within an integrated water resource management approach.

 However, concern about the risk to public health makes wastewater use a controversial issue and may limit its planned extension (see Chapter 5). Guidelines on wastewater use in agriculture typically stipulate treatment levels and processes, although in 2006 the World Health Organization (WHO) published guidelines that utilize a risk management approach and recommend the introduction of barriers to risks along the pathway from wastewater production to crop consumption (WHO, 2006). This offers a pragmatic and workable solution that is designed to protect farmers as well as consumers.

Sustainable Crop Selection for Drylands

Because of the extreme aridity of many of the areas under discussion, water is most efficiently used on plants that can become multifunctional features in the landscape, as part of a new ecosystem. Every plant is then a multi-purpose species, capable of breaking the force of the wind and absorbing water, but also of producing food, fruit, oil, fodder and firewood, and of fixing nitrogen, hosting useful or edible insects, or providing building material, hence providing a multitude of ecosystem services. Cultivating a diverse and sustainable crop repertoire would support livelihoods at the local level by making dryland agroecosystems more productive and sustainable, thus increasing the amount of usable materials for the desert household.

An investigation of local plants in each candidate site helps to identify suitable plants, i.e. which local plants may be a valuable source of food for the human population, which can be utilized to support the flocks and herds, and those that may be necessary for the restoration of water and nutrient cycling in the most degraded areas (Bainbridge, 2004). Many suitable crops might be found among the local perennial plants (Shmida and Darom, 1992). Perennial plants and their longer cycles of living and yielding are much more suitable to the desert than annual or seasonal crops as they need little tillage and are more water thrifty. Being adapted to the slow breakdown of organic matter and release of minerals in dryland soils, such perennials allow for natural regeneration of soil structure, while each litre of water invested in a perennial is converted to long-lived plant tissue, fruit, seeds and leaves (Solowey, 2010).

As an example, Table 6.1 lists several local crop candidates from a zone of hyper-aridity shared by Israel and Jordan, which can be grown in areas with 50 to 120 mm of rainfall utilizing water harvesting technologies. Some of these plants were introduced from other drylands to Israel and Jordan through cooperative programmes between the Arava Institute for Environmental Studies and the Jordan University of Science and Technology. Others are wild plants undergoing an accelerated process of domestication. The advantages of using such desert-adapted plants

include the water-thrifty nature of the germplasm, the availability of fresh genetic material with no need for quarantine, local knowledge, and familiarity relative to the plant material and possibly existing systems for utilization of the plant products. All plants in Table 6.1 are physiologically appropriate for arid and hyper-arid areas, i.e. drought-resistant, and multipurpose, i.e. producing food and material for sale and trade. The plants were selected because of their tolerance for high pH soil and their physical influences in various cultivation formats, which enable them to improve the organic matter content of poor soils and soil permeability. Their medicinal value and their value as browse and feedstock were also taken into account. Many of these plants could support small-scale value-added product manufacture.

Perennial plantations, which ideally are made up of various species, such as in most oases, are regeneration friendly. They may make best use of the available water and may help to generate supporting and regulatory ecosystem services. Trees shade and protect the soil from the sun, lowering soil temperatures and thereby regulating the microclimate. Fallen leaves produce natural mulch and encourage colonization by beneficial soil organisms. Trees and perennial plants are sanctuaries and nesting places for birds, hunting grounds for insectivores and feeding areas for pollinating insects. Their roots are highways into the earth for ants, beneficial nematodes, beneficial fungi and mycorrhizae, as well as conduits for sparse and precious rainfall.

When perennial plantations are established, their mitigating presence allows for the integration of some annual plants to utilize the runoff from irregular rains. Perennial trees can thus be combined with annual elements to enhance biodiversity and provide multiple benefits (Solowey, 2010). The annual plants may include grass for grazing, medicinal herbs for personal use or cottage industry, and leafy vegetables to improve the diet of the farmer and herder. Hence, a balanced agroecosystem can be established, with a wealth of regulatory and supporting ecosystem services, safeguarding the delivery of food and other provisioning services. In semi-arid areas, well-managed rangelands or arboreal pastures

Table 6.1. Crop candidates and the potential contributions of their germplasm to ecosystem services in a desert area shared by Israel and Jordan. Plants not native to Israel and Jordan are in bold.

Crop candidate	Provisioning services	Regulatory services[a]	Supporting services[b]
Acacia	Sap, pods, wood, browse	Xerophyte, apiary	Pioneer
Achillea	Essential oil, flowers, medicinal	Apiary	Pioneer
Argania spinosa	Nuts, oil, wood, poles, browse	Apiary	Reclamation
Artemisia spp.	Essential oil for medicinal use (antimalarial)	Apiary	
Atriplex spp.	Flowers, pasture, medicinal	Apiary	
Balanites spp.	Fruit, oil, flowers, sap, leaves, medicinal, poles, fence, browse	Shade	
***Boswellia* spp.**	Sap, incense, wood for smoking, medicinal	Apiary	Reclamation
***Capparis spinosa* (capers)**	Buds, medicine, cosmetics, liquor	Apiary	Ground cover
Cassia spp.	Flowers, leaves, pods	Apiary	Ground cover, reclamation
***Commiphora* spp.**	Sap, wood for smoking, flowers, medicinal	Xerophyte, apiary	Reclamation
Haloxylon spp. (saxaul)	Browse, sap, flowers	Dune stabilization	Reclamation
Pistacia terebinthus (terebinth)	Resin, wood, browse, rootstocks	Shade, windbreak	Reclamation
***Prosopis* spp.**	Browse, wood, poles, pods	Stabilization, apiary, windbreak	Reclamation
Salicornia spp. (glasswort)	Browse, flowers, oil	Apiary	Reclamation, pioneer
***Sclerocarya birrea* (marula)**	Fruit, oil, timber, liquor, browse	Shade	Reclamation
Ziziphus spp.	Fruit, poles, liquor, juice, browse	Living fence, windbreak	
Zygophyllum spp.	Browse, sap, flowers, pasture, medicinal		Pioneer

[a] Apiary plants are important habitats for bees, and hence contribute to pollination. Shade plants and windbreaks play a role in climate regulation. Xerophytes use very little water so help to regulate water flows. Living fence and (dune) stabilization are important in erosion regulation.
[b] Reclamation plants and ground cover help soil formation and nutrient cycling. Pioneer plants contribute to the mitigation of climate change.

could have similar impacts (see below). The perennial trees would ideally be multi-purpose, providing fruits, shade, fodder, wood and more. A good example of such a multi-purpose tree is the lalob – one of many common names (*Balanites aegyptiaca*), which supplies browse for goats and camels, fruit pulp for fermentation, medicinal sap, oil of good quality for illumination and firewood; it can also serve as an anti-erosion plant (National Research Council, 2008).

Another interesting example of a multipurpose tree is the argan (*Argania spinosa*) of southern Morocco, which produces hardwood for tool manufacture when coppiced, can be a source of browsing for goats, a source of nectar and pollen for honey bees and an anti-erosive tree in areas with seasonal flooding but, most of all, is a source of edible oil, soap and cosmetic oil for the local people. For example, argan oil is used very much like olive oil in the Moroccan kitchen; it is also added to a porridge-like dish (semetar), and the roasted nuts are used to make argan nut butter (amalou) after extraction of the oil (Morton, 1987). Internationally, argan oil has become increasingly popular for cosmetic use on skin and hair, and there are claims that it benefits local livelihoods as well as the environment. While it does indeed seem to have a povertyreducing impact and aid increased access to education for girls, the argan forest itself may now be under even more threat than it was before (Lybbert *et al.*, 2011).

Members of the *Prosopis* family of trees are all nitrogen fixers, as well as being multi-purpose trees. These trees can supply browse, high-quality protein food from pods, firewood, syrup and non-gluten flour for human consumption, shade and shelter for flocks, and building materials; they can also be used as windbreaks (Knoop *et al.*, 2012). In the dry season especially, the trees provide high-quality feed for livestock. Unfortunately, several *Prosopis* species have a tendency to invasiveness that needs to be carefully managed; they also need to be thinned to allow for the planting or emergence of other species. Introduced species should always be evaluated for weedy properties (Solowey, 2003).

Grazing in Drylands

Dry rangelands support about 50% of the world's livestock population (Millennium Ecosystem Assessment, 2005) and are of huge importance for the often poor livestock keepers in these regions. The most important livestock production systems in dry areas are grazing systems, which occupy 77% of the dryland area worldwide; these are followed by mixed rainfed systems, with a share of 17%. Livestockdominated and mixed crop–livestock systems in drylands cover about 11.9 and 6.9 million km², respectively, or about 15% and 9% of the 80.8 million km² comprising Latin America, Africa and South and South-east Asia (Thornton *et al.*, 2002; Table 6.2). In 2002, livestockdominated areas were home to about 116 million people, whereas about 595 million people resided in mixed crop–livestock systems.

Table 6.2. Distribution of land and people in mixed crop–livestock and livestock dominated systems in drylands in developing countries (based on Thornton *et al.*, 2002).

	Livestock-dominated systems	Mixed crop–livestock systems
Land area (million km²)	11.9	6.9
Land area as % of country	15	9
Number of people (million)	116	595
Density (people/km²)	9.7	86.2

Herding can be viewed as a form of water harvesting in the sense that grazing animals capture the benefits of sparsely distributed rainfall by grazing pastures (Bindraban *et al.*, 2010). Mobility is the primary and requisite characteristic of pastoral agroecosystems. Grazing by domestic and wild ungulates is the means of maintaining extensive grasslands that provide important ecosystem services, including the maintenance of biodiversity and carbon sequestration. At the same time, extensive cattle enterprises have been responsible for 65–80% of the deforestation of the Amazon at a rate of forest loss of 18–24 million ha/year (Herrero *et al.*, 2009).

In recent decades, the expansion of cultivation and the establishment of international boundaries and barriers across traditional migratory routes have diminished mobility, forcing herders towards a more sedentary livelihood strategy that has often resulted in severe land and water degradation, aggravated poverty, poor health and food insecurity. The importance of rangelands for livestock grazing is highest in the arid agroecosystems, whereas in the semi-arid and sub-humid areas, grasslands are being converted into shrublands and cultivated land (Millennium Ecosystem Assessment, 2005). Small areas of encroaching cultivation can have a multiplier effect and reduce livestock production over much larger land areas. In arid regions, the expansion of cropland, inappropriate grazing practices (Geist and Lambin, 2004) and newly imposed barriers to the mobility of pastoralists may even increase trends in desertification. Policies directed to making nomads sedentary often have adverse effects as they reduce the traditional ability of pastoralists to respond to climate shocks, resulting in a downward spiral of poverty, conflict and social exclusion (de Jode, 2010).

In the tropics, the expansion of croplands at the expense of grazing areas is driven by increasing human populations (Kristjanson *et al.*, 2004). As a result, in the sub-humid and semi-arid tropics, traditional pastoral practices are often being replaced with agropastoralism and mixed farming in which livestock increasingly depend on crop residues as feed. The transition from grazing to agropastoralism to mixed crop–livestock production is often

accompanied by the migration of people, and an increased human population also puts enhanced pressure on fuel sources such as charcoal, further aggravating land degradation. Increased migration of people may lead to conflicts over access to natural resources, such as water resources, that are used by livestock keepers for drinking but also claimed by farmers for irrigating their crops. However, the increased interaction between pastoralists and farmers may lead to increased exchanges and closer collaboration too (Turner, 2004).

Inappropriate livestock grazing practices are often seen as the culprit causing rangeland degradation and desertification (Asner *et al.*, 2004). Traditional pastoral practices are generally well adapted to make use of the spatially and temporally variable feed resources in rangelands (IIDE and SOS Sahel UK, 2010), but when these are disrupted or pressured as a result of demographic, climate or land use changes, livestock grazing may threaten the provision of ecosystem services. Overgrazing is a leading cause of land degradation in arid drylands, tropical grasslands and savannas worldwide. It leads to soil compaction, reduction in long-term grazing productivity, loss of topsoil, disruption of the hydrological cycle and deterioration of water quality. In such degraded rangelands, most water is lost as runoff and unproductive evaporation, so that water use efficiency is dramatically reduced. Increased runoff and the trampling of the soil by livestock lead to erosion and thence to siltation of downstream freshwater resources. This may lead to soil and vegetation degradation, reduced productivity and, eventually, food insecurity (Asner *et al.*, 2004).

Although reports from drylands often paint grim pictures of poverty, drought and conflicts over resources, the degradation of drylands could be avoided by intensifying agricultural production and safeguarding pastoral mobility (Millennium Ecosystem Assessment, 2005). Options for carbon storage could be enhanced, as, because of their large area, rangelands could be a global sink of a roughly similar size to forests (Herrero *et al.*, 2009; Box 2.3, Chapter 2). There is a real need for research on how this large potential can be tapped through technologies and policies for carbon sequestration. Rangelands could even be the source of significant regional increases in water productivity by judiciously using them as a feed source, at the same time as taking care to avoid overgrazing (Herrero *et al.*, 2009; Bindraban *et al.*, 2010).

Solutions for breaking the downward spiral of over-exploitation, degradation and disrupted ecosystem services in drylands should take on board the technical, sociopolitical and institutional issues that are involved (Amede *et al.*, 2009). Such solutions should secure property rights, be risk averse and take into account the labour constraints of women, men and children, as well as enabling their access to input and output value chains and market information (FAO *et al.*, 2010). In particular, securing the mobility of herds for access to natural resources, trade routes and markets is essential to avoid degradation and conflict (de Jode, 2010). This can be achieved through appropriate policies that take into account transboundary herd movements but also enable the creation of corridors and the establishment of water points and resting areas along routes. The strategic positioning of drinking water points helps to avoid the concentration of too many animals around one watering point, which would cause soil and vegetation degradation and water contamination (Brits *et al.*, 2002; Wilson 2007), and is instrumental in balancing feed availability with livestock numbers so that feed resources can be used optimally (Peden *et al.*, 2009).

Rangelands can be improved by changing them into arboreal pastures, using appropriate multifunctional perennial and annual species. The animals play their own role in the establishment, survival and distribution of plant species. Most herbivores prefer soft, fast-growing plants, so these disappear first. More resinous, nasty-tasting, spiny or tough plants – often the typical desert species – are eaten more slowly. Thus, grazing animals have their own impact on water availability, with wild herbivore populations fluctuating dramatically in response to rain and vegetation (Bainbridge, 2007), whereas domesticated livestock can survive and maintain herd size by feeding on cultivated perennial grasses and trees. Trees in grazing areas help to mitigate the impacts of the wind and water erosion that are rampant in degraded drylands. Planting arboreal pastures

may also reduce competition between local animal herders over the rights to graze sheep and goats on the little bits of remaining vegetation.

Arboreal pastures are often suitable for reclaiming and using the wastewater and runoff that are currently damaging factors and turning them into water resources for the deliberate increase of native vegetation, so that the land degradation process is reversed. The ultimate goal is an increase of vegetation for grazing, the establishment of partnerships for sustainable grazing sites between former rivals and, it is hoped, the creation of examples that can be emulated in other contested, arid and desolate grazing areas to the benefit of all stakeholders (Evenari *et al.*, 1982). Arboreal pasturage can, therefore, provide a wide range of ecosystem services in addition to grazing grounds, such as erosion control and enrichment of the soil by leaf litter and from the nitrogen fixation that is done by appropriate tree species (Rabia *et al.*, 2008).

Conclusions

Agriculture and pastoralism in drylands are challenged by the scarcity of various natural resources, in particular water and soil fertility. These conditions require site-specific solutions that include seeking synergies between agriculture and anti-desertification efforts. Degraded lands could be brought back under productive use through rangeland conservation and better farming practices, which, in turn, restore surface vegetation and soil functions, in particular water retention. New technologies, new cultivars and enhanced utilization of water resources can thus be combined to strengthen ecosystem services and increase water efficiency for the cultivation of suitable crops and modified rangelands. Strategies for more sustainable models of arid land agriculture include the efficient collection of runoff, soil-based storage of moisture and nutrients, and strategic planting of local and desert-adapted cultivars to increase the resource base and the provision of ecosystem services. When combined with organic fertilizers that increase the water-holding capacity of the soil, effective weed control and crop protection against pest and diseases, productivity under semi-arid conditions can triple or more and hence make more effective use of rainwater (Bindraban *et al.*, 1999).

In semi-arid rangelands, providing incentives to livestock herders can help to improve herd management and safeguard the regulation and support of ecosystem services. These strategies must take into account the differentiated needs and capacities of local men and women and of different social groups, in this manner ensuring that those responsible for certain tasks are effectively able to accomplish them.

Such approaches are not necessarily technically complex, but they do require a wholesale shift towards more integrated approaches to agroecosystem management, building on the common goal of sustainability.

They also require the building of institutional capacity and collective action to facilitate adoption and dissemination of these good practices in drylands. Looking at water, ecosystem and human needs in parallel, and identifying and building upon mutually supportive approaches, is the key, as is looking across sectors. By linking and combining appropriate production systems in a landscape, synergies can be explored. Drought-resistant plants, arboreal pastures and perennial grasses can be cultivated in a landscape with strategically placed corridors and water points for herds, thereby providing more sustainable exploitation options for agropastoralists. The integration of crop, tree, livestock and, in some cases, aquaculture, can enhance resource recovery and the reuse of resources for feed or soil fertility.

References

Amede, T., Geheb, K. and Douthwaite, B. (2009) Enabling the uptake of livestock-water productivity interventions in the crop–livestock systems of sub-Saharan Africa. *The Rangeland Journal* 31, 223–230. doi:10.1071/RJ09008

Amede, T., Tarawali, S. and Peden, D. (2011) Improving water productivity in crop livestock systems of drought prone regions. Editorial Comment. *Experimental Agriculture* 47, 1–6.

Asner, G.P., Elmore, A.J., Olander, L.P., Martin, R.E. and Harris, A.T. (2004) Grazing systems, ecosystem responses and global change. *Annual Review of Environment and Resources* 29, 261–299. doi:10.1146/annurev.energy.29.062403.102142

Bainbridge, D.A. (2004) The anatomy, physiology, psychology and economics of desert destruction and restoration. Paper presented to: *3rd Mojave Desert Science Symposium, November 16–18, 2004, University of Redlands, California.*

Bainbridge, D.A. (2007) *A Guide for Desert and Dryland Restoration.* Island Press, Washington, DC.

Behnke, R.H. Jr, Scoones, I. and Kerven, C. (eds) (1993) *Range Ecology at Disequilibrium: New Models of Natural Variability and Pastoral Adaptation in African Savannas.* Overseas Development Institute, London.

Bindraban, P.S., Verhagen, A., Uithol, P.W.J. and Henstra, P. (1999) *A Land Quality Indicator for Sustainable Land Management: The Yield Gap.* Report 106, Research Institute for Agrobiology and Soil Fertility (AB-DLO), Wageningen, the Netherlands.

Bindraban, P., Conijn, S., Jongschaap, R., Qi, J., Hanjra, M., Kijne, J., Steduto, P., Udo, H., Oweis, T. and de Boer, I. (eds) (2010) *Enhancing Use of Rainwater for Meat Production on Grasslands: An Ecological Opportunity Towards Food Security.* Proceedings 686, International Fertiliser Society, Cambridge, 10 December 2010, Leek, UK. ISRIC World Soil Information, Wageningen, the Netherlands.

Brits, J., van Rooyen, M.W. and van Rooyen, N. (2002) Ecological impact of large herbivores on the woody vegetation at selected watering points on the eastern basaltic soils in the Kruger National Park. *African Journal of Ecology* 40, 53–60. doi:10.1046/j.0141-6707.2001.00344.x

Burke, E.J., Brown, S.J. and Christidis, N. (2006) Modeling the recent evolution of global drought and projections for the twenty-first century with the Hadley Centre climate model. *Journal of Hydrometeorology* 7, 1113–1125. doi:10.1175/JHM544.1

de Jode, H. (ed.) (2010) *Modern and Mobile. The Future of Livestock Production in Africa's Drylands.* International Institute for Environment and Development (IIED), London and SOS Sahel International UK, Oxford, UK.

Drechsel, P., Cofie, O.O., Keraita, B., Amoah, P., Evans, A. and Amerasinghe, P. (2011) Recovery and reuse of resources: enhancing urban resilience in low-income countries. *Urban Agriculture Magazine* 25, 66–69.

Ellis, J.E. and Swift, D.M. (1988) Stability of African pastoral ecosystems: alternate paradigms and implications for development. *Journal of Range Management* 41, 450–459.

Evenari, M., Shanan, L. and Tadmor, N. (1982) *Negev: The Challenge of a Desert*, 2nd edn. Harvard University Press, Cambridge, Massachusetts.

FAO (2013) Municipal wastewater: from production to use (with link to the AQUASTAT database). Food and Agriculture Organization of the United Nations, Rome. Available at: http://www.fao.org/nr/water/aquastat/wastewater/index.stm (accessed February 2013).

FAO, IFAD and ILO (2010) *Gender Dimensions of Agricultural and Rural Employment: Differentiated Pathways out of Poverty. Status, Trends and Gaps.* Food and Agriculture Organization of the United Nations, Rome, International Fund for Agricultural Development, Rome and International Labour Organization, Geneva, Switzerland.

Folke, C. (2006) Resilience: the emergence of a perspective for social-ecological systems analysis. *Global Environmental Change* 16, 253–267. doi:10.1016/j.gloenvcha.2006.04.002

Geist, H.J. and Lambin, E.F. (2004) Dynamic causal patterns of desertification. *BioScience* 54, 817–829.

GIAHS (2013) Globally Important Agricultural Heritage Systems, Available at: http://www.giahs.org (accessed February 2013).

Hamilton, A.J., Stagnitti, F., Xiong, X., Kreidl, S.L., Benke, K.K. and Maher, P. (2007) Wastewater irrigation: the state of play. *Vadose Zone Journal* 6, 823–840.

Herrero, M., Thornton, P.K., Gerber, P. and Reid, R.S. (2009) Livestock, livelihoods and the environment: understanding the trade-offs. *Current Opinion in Environmental Sustainability* 1, 111–120. doi:10.1016/j.cosust.2009.10.003

Herrmann, S.M., Anyamba, A. and Tucker, C.J. (2005) Recent trends in vegetation dynamics in the African Sahel and their relationship to climate. *Global Environmental Change* 15, 394–404. doi:10.1016/j.gloenvcha.2005.08.004

Jiménez, B., Drechsel, P., Koné, D., Bahri, A., Raschid-Sally, L. and Qadir, M. (2010) Wastewater, sludge and excreta use in developing countries: an overview. In: Drechsel, P., Scott, C.A., Raschid-Sally, L., Redwood, M. and Bahri, A. (eds) (2010) *Wastewater Irrigation and Health: Assessing and Mitigating Risk in Low-income Countries.* Earthscan, London, International Water Management Institute (IWMI), Colombo and International Development Research Centre (IDRC), Ottawa, pp. 3–27.

Johnston, R. and McCartney, M. (2010) *Inventory of Water Storage Types in the Blue Nile and Volta River Basins.* IWMI Working Paper 140, International Water Management Institute, Colombo. doi: 10.5337/2010.214

Kirkby, J., O'Keefe, P. and Timberlake, L. (1995) *The Earthscan Reader in Sustainable Development.* Earthscan, London.

Knoop, L., Sambalino, F. and Van Steenbergen, F. (2012) *Securing Water and Land in the Tana Basin: A Resource Book for Water Managers and Practitioners.* 3R Water Secretariat, Wageningen, the Netherlands.

Kristjanson, P.M., Thornton, P.K., Kruska, R.L., Reid, R.S., Henninger, N., Williams, T.O., Tarawali, S.A., Niezen, J. and Hiernaux, P. (2004) Mapping livestock systems and changes to 2050: implications for West Africa. In: Williams, T.O., Tarawali, S.A., Hiernaux, P. and Fernández-Rivera, S. (eds) *Sustainable Crop-livestock Production for Improved Livelihoods and Natural Resource Management in West Africa. Proceedings of an International Conference Held at International Institute of Tropical Agriculture (IITA), Ibadan, Nigeria, 19–22 November 2001.* International Livestock Research Institute (ILRI), Nairobi and Technical Centre for Agricultural and Rural Cooperation, ACP-EC (CTA), Wageningen, the Netherlands, pp. 28–44. Available at: http://ilri.org/InfoServ/Webpub/fulldocs/SustainableCropLivestock/Pg028_044%20Kristjanson%20and%20Thornton.pdf (accessed February 2013).

Lybbert, T.J., Aboudrare, A., Chaloud, D., Magnan, N. and Nash M. (2011) Booming markets for Moroccan argan oil appear to benefit some rural households while threatening the endemic argan forest. *Proceedings of the National Academy of Sciences of the United States of America* 108, 13963–13968. doi:10.1073/pnas.1106382108

McCartney, M. and Smakhtin, V. (2010) *Water Storage in an Era of Climate Change: Addressing the Challenge of Increasing Rainfall Variability.* IWMI Blue Paper, International Water Management Institute, Colombo.

Millennium Ecosystem Assessment (2005) *Ecosystems and Human Well-being: Synthesis. A Report of the Millennium Ecosystem Assessment.* World Resources Institute and Island Press, Washington, DC. Available at: www.maweb.org/documents/document.356.aspx.pdf (accessed February 2013).

Morton, J.F. (1987) *Fruits for Warm Climates.* Florida Flair Books, Miami, Florida.

Mugerwa, S. (2009) Effect of reseeding and cattle manure on pasture and livestock water productivity in rangelands of Nakasongola District, Uganda. MSc Thesis, Makerere University, Kampala.

National Research Council (2008) *Lost Crops of Africa: Volume III: Fruits.* National Academies Press, Washington, DC.

Ngigi, S.N. (2003) Rainwater harvesting for improved food security: promising technologies in the Greater Horn of Africa. Greater Horn of Africa Rainwater Partnership and the Kenya Rainwater Association, Nairobi, Kenya.

Olsson, L., Eklundh, L. and Ardö, J. (2005) A recent greening of the Sahel: trends, patterns and potential causes. *Journal of Arid Environments* 63 (3), 556–566. doi:10.1016/j.jaridenv.2005.03.008

Peden, D., Taddesse, G. and Haileslassie, A. (2009) Livestock water productivity: implications for sub-Saharan Africa. *The Rangeland Journal* 31, 187–193. doi:10.1071/RJ09002

Rabia, A.R., Solowey, E. and Leu, S. (2008) Environmental and economic potential of Bedouin dryland agriculture: a case study in the Northern Negev, Israel. *Journal of Management of Environmental Quality* 19, 353–366.

Reij, C., Tappan, G. and Belemvire, A. (2005) Changing land management practices and vegetation on the Central Plateau of Burkina Faso (1968–2002). *Journal of Arid Environments* 63, 642–659. doi:10.1016/j.jaridenv.2005.03.010

Reij, C., Tappan, G. and Smale, M. (2009) *Agroenvironmental Transformation in the Sahel: Another Kind of "Green Revolution".* IFPRI Discussion Paper 00914, 2020 Vision Initiative. Prepared for the project on Millions Fed: Proven Successes in Agricultural Development, International Food Policy Research Institute (IFPRI), Washington, DC. Available at: www.ifpri.org/sites/default/files/publications/ifpridp00914.pdf (accessed December 2012).

Scheffer, M., Carpenter, S.R., Foley, J.A., Folke, C. and Walker, B. (2001) Catastrophic shifts in ecosystems. *Nature* 413, 591–596.

Scott, C., Drechsel, P., Raschid-Sally, L., Bahri, A., Mara, D., Redwood, M. and Jiménez, B. (2010) Wastewater irrigation and health: challenges and outlook for mitigating risks in low-income countries. In: Drechsel, P., Scott, C.A., Raschid-Sally, L., Redwood, M. and Bahri, A. (eds) *Wastewater Irrigation and Health: Assessing and Mitigating Risk in Low-income Countries.* Earthscan, London, International Water Management Institute, Colombo and International Development Research Centre, Ottawa, pp. 381–394.

Shmida, A. and Darom, D. (1992) *Handbook of Trees and Bushes of Israel.* Keter Publishing House, Jerusalem [in Hebrew].

Solowey, E. (2003) Appendix four, characteristics associated with weediness. In: Solowey, E. *Small Steps Towards Abundance: Crops for a More Sustainable Agriculture.* Biblio Books Israel, Acco, Israel/Biblio Books International, Miami, Florida.

Solowey, E. (2010) Chapter 7. Arboreal pastures. In: Solowey, E. *Growing Bread on Trees: The Case for Perennial Agriculture.* Biblio Books Israel, Acco, Israel/Biblio Books International, Miami, Florida, pp. 106–123.

Stringham, T.K., Krueger, W.C. and Shaver, P.L. (2003) State and transition modeling: a process based approach. *Journal of Range Management* 56, 106–113.

Thornton, P.K., Kruska, R.L., Henninger, N., Kristjanson, P.M., Reid, R.S., Atieno, F., Odero, A.N. and Ndegwa, T. (2002) *Mapping Poverty and Livestock in the Developing World.* International Livestock Research Institute, Nairobi.

Turner, M.D. (2004) Political ecology and the moral dimensions of "resource conflicts": the case of farmer–herder conflicts in the Sahel. *Political Geography* 23, 863–889.

UNCCD (2010) *United Nations Decade for Deserts and the Fight Against Desertification.* United Nations Convention to Combat Desertification, Bonn, Germany.

Vohland, K. and Boubacar, B. (2009) A review of *in situ* rainwater harvesting (RWH) practices modifying landscape functions in African drylands. *Agriculture, Ecosystems and Environment* 131, 119–127. doi:10.1016/j.agee.2009.01.010

Washington-Allen, R.A., and Salo, L.F. (2007) Meeting review: catastrophic thresholds, perspectives, definitions, and applications. *Bulletin of the Ecological Society of America* 88, 219–225. Available at: http://www.esajournals.org/doi/full/10.1890/0012-9623(2007)88%5B219%3ACTPDAA%5D2.0.CO%3B2.

Washington-Allen, R.A., Ramsey, R.D., West, N.E. and Norton, B.E. (2008) Quantification of the ecological resilience of drylands using digital remote sensing. *Ecology and Society* 13(1): 33. Available at: http://www.ecologyandsociety.org/vol13/iss1/art33/ (accessed February 2013).

Weckenbrock, P., Evans, A.E.V., Qaiser Majeed, M., Ahmad, W., Bashir, N. and Drescher, A. (2011) Fighting for the right to use wastewater: what drives the use of untreated wastewater in a peri-urban village of Faisalabad, Pakistan? *Water International* 36, 522–534.

WHO (2006) *Guidelines for the Safe Use of Wastewater, Excreta and Greywater in Agriculture.* World Health Organization, Geneva, Switzerland.

Wilson, T. (2007). Perceptions, practices, principles and policies in provision of livestock water in Africa. *Agricultural Water Management* 90, 1–12. doi:10.1016/j.agwat.2007.03.003

Winpenny, J., Heinz, I. and Koo-Oshima, S. (2010) *The Wealth of Waste: The Economics of Wastewater Use in Agriculture.* FAO Water Reports 35, Food and Agriculture Organization of the United Nations, Rome.

7 Wetlands

Max Finlayson,[1*] Stuart W. Bunting,[2†] Malcolm Beveridge,[3] Rebecca E. Tharme[4] and Sophie Nguyen-Khoa[5]

[1]Institute for Land, Water and Society (ILWS), Charles Sturt University, Albury, New South Wales, Australia; [2]Essex Sustainability Institute, University of Essex, Colchester, UK; [3]WorldFish, Lusaka, Zambia; [4]The Nature Conservancy (TNC), Buxton, UK; [5]World Water Council (WWC), Marseille, France

Abstract

After commencing with a summary of the current status, importance and productivity of natural wetlands, the chapter reviews the contribution of wetland ecological functions to sustaining vital ecosystem services. Wetlands are vulnerable to a range of anthropogenic pressures, notably land use change, disruption to regional hydrological regimes as a result of abstraction and impoundment, pollution and excessive nutrient loading, the introduction of invasive species and overexploitation of biomass, plants and animals. Natural wetlands have often been modified to accommodate agricultural and aquaculture production, or wetlands may be created in the process of establishing farming systems. Prospects for established practices, such as culturing fish in rice fields, culture-based fisheries and integrating aquaculture with livestock production or into water storage and irrigation schemes are critically reviewed. Apparent conflicts between agricultural development and intensification and wetland conservation are discussed, and opportunities to reconcile competing demands are considered. Wetlands, whether classified as natural or as agroecosystems, sustain a wide range of ecosystem services that contribute to water and food security, but the appropriation of these services should be maintained with adequate provision for sustaining environmental stocks and flows and conserving and protecting aquatic biodiversity.

Background

Globally, wetlands[1] cover at least 6% of the earth's terrestrial surface (Finlayson and D'Cruz, 2005), of which substantively 200–280 million ha occur in Asia, followed by 125–130 million ha in Africa (Table 7.1).

Common inland and coastal wetlands comprise lakes, rivers, marshlands, mangroves, estuaries and lagoons, and aquifer systems, through to shallow water coral reefs and seagrass beds. These ecosystems host a wealth of biodiversity and arguably account for about 45% of the total economic value of all global ecosystem

---

* E-mail: mfinlayson@csu.edu.au
† E-mail: swbunt@essex.ac.uk

Table 7.1. Estimates of global wetland area for the six geopolitical regions used by the Ramsar Convention on Wetlands (Ramsar Convention Secretariat, 2011).

	Estimates of global wetland area (million ha and percentage area)	
Region	Global lakes and wetlands database (Lehner and Döll, 2004)	Global review of wetland resources (Finlayson *et al.*,1999)
Africa	131 (14%)	125 (10%)
Asia	286 (32%)	204 (16%)
Europe	26 (3%)	258 (20%)
Latin America	159 (17%)	415 (32%)
North America	287 (31%)	242 (19%)
Oceania	28 (3%)	36 (3%)
Total	917 (100%)	1,280 (100%)

services, although estimates vary (Millennium Ecosystem Assessment, 2005; see also the discussion on the valuing of ecosystem services in Chapters 3 and 4). The supply of fresh water to human populations is recognized as one of the foremost natural benefits of wetlands, coupled with the provision of those services that support food security and reduce rural poverty, such as capture fisheries and sustainable aquaculture (Millennium Ecosystem Assessment, 2005; Dugan *et al.*, 2007). In many instances, though, the relative contributions of different wetlands types towards food production and food security have not been determined, or are highly variable, as found for wetlands in sub-Saharan Africa (Rebelo *et al.*, 2010, 2011; McCartney *et al.*, 2011b). Other important benefits associated with wetlands include base-flow releases during dry seasons, the capacity to provide off-season biomass (fish and crops) and their role as biodiversity hotspots – they often provide habitats for nationally or globally threatened species, though once again, the evidence base for all such benefits may not be that strong (e.g. McCartney *et al.*, 2011a).

Wetland Ecosystem Services

For this book, the focus on wetlands is their role within the hydrological cycle, where they contribute towards a complex series of hydrological regulatory functions, including water storage (i.e. water holding, groundwater recharge and discharge, and flood prevention or attenuation by flow regulation and mitigation), water purification and the retention of nutrients and sediments (Millennium Ecosystem Assessment, 2005; Finlayson, 2011; Chapter 3). The quantity of water stored globally in wetlands amounts to about 11.5 thousand km^3 (Shiklomanov and Rodda, 2003). It is important to note that most of this is cycled through different wetlands. The elimination of wetlands, and thus the need to maintain hydrological flows to them, may be seen by some as freeing up water for human appropriation but, generally, it reduces the availability of water for direct human use.

Wetlands, notably river floodplains and some upper catchment palustrine wetlands (e.g. in the Andean páramo), are often regarded as functioning as natural sponges; they expand to accommodate excess water in times of heavy rain and contract as they release water slowly throughout the dry season, thereby maintaining streamflow (Millennium Ecosystem Assessment, 2005). In reality, the hydrological functions of most wetlands are more complex and vary considerably among sites (Bullock and Acreman, 2003; McCartney *et al.*, 2010, 2011a). Inland wetlands, in particular, play a major role in providing water for agriculture (Falkenmark *et al.*, 2007). For example, the Hadejia-Nguru wetlands in northern Nigeria play a major role in recharging aquifers that provide domestic water supplies to approximately a million people (Hollis *et al.*, 1993).

The flood mitigation services of wetlands are particularly valuable, especially where they reduce flood risks to housing, industry and infrastructure. Policy and public sentiment in many countries is moving away from artificial flood control approaches (e.g. embankments) towards wetland rehabilitation because it is often cheaper and more sustainable. Male and female farmers are often integral to this process, either because they too have an interest in better flood protection of their assets, or through the receipt of incentives (compensation) from urban areas for reinstating flood protection on their farmlands and reverting to more traditional floodplain pasture cropping or grazing. In New Zealand, formal protection of the Whangamarino Wetland led to reduced costs for flood protection, while conserving water for irrigation during the dry season (Department of Conservation, 2007).

Natural wetlands have often been modified to accommodate agricultural and aquacultural production. Wetlands may also be created in the process of establishing farming systems in the form of storage reservoirs and fish ponds, for example; the resulting array of managed aquatic ecosystems are referred to collectively here as wetland agroecosystems. With agricultural expansion into wetlands, and the growing need to produce more food with less water, it is important that the functions of these agroecosystems are seriously considered and managed in terms of their contributions to ecosystem services (Falkenmark et al., 2007; Wood and van Halsema, 2008). Key ecological attributes or functions of wetlands, including sediment and nutrient transport and delivery into estuaries or on to river floodplains, generally enhance food production in downstream agroecosystems. Another important function of wetland agroecosystems is the treatment of wastes. This is facilitated by physical, biological and biochemical processes, but there are intrinsic limits to the waste-processing capabilities of wetlands. Aquatic ecosystems assimilate on average 80% of the global nitrogen load, but this intrinsic self-purification capacity varies widely and is declining as a result of the loss of wetland areas and overloading of the self-purification capacity (Millennium Ecosystem Assessment, 2005; Deegan et al., 2012).

The provision of ecosystem services by wetlands is often undervalued and assumed only to comprise fish catches. However, a wide array of other aquatic animals and plants from wetlands are exploited by various groups of people at various times, and often by the poor in times of need (WRI et al., 2008). Artificial water bodies and wetland agroecosystems also sustain a range of provisioning ecosystem services and, with the proliferation of water storage reservoirs for irrigation and electricity generation, are emerging as a major source of food and income in remote and highland areas (Welcomme et al., 2010).

The most common wetland agroecosystems are rice fields, the total area of which exceeds 125 million ha, and covers some 9% of the earth's arable land (Maclean et al., 2002). These continue to provide employment and staple food supplies for a large proportion of the rural poor in Asia. Of the total area planted with rice, just over half (55%) has been estimated to be under irrigation (Frei and Becker, 2005). These vital wetland agro-ecosystems support a wide range of biodiversity, including fish, amphibians and insects, and can play a significant role in the conservation of waterbird populations (Matsuno et al., 2002; Bellio et al., 2009; Elphick et al., 2010). The collection of fish and other aquatic animals by farming households and local communities for food and to sell can often constitute a major benefit of having access to inundated rice fields (Amilhat et al., 2009; see subsection below on 'Aquaculture in rice fields').

Such benefits may not be realized in intensively managed rice fields where the natural water regime has been altered and pesticide use is routine. Still, these fields also provide natural drainage systems and help in flood control, although in circumstances where wetlands have been converted to rice fields, there is little information about whether these benefits have been enhanced or have declined. There is also evidence that the construction of rice fields does not substitute for the biodiversity values that were previously obtained from lost or altered wetlands (Bellio et al., 2009; Elphick et al., 2010). Methane and nitrous oxide emissions from flooded rice fields are a significant source of anthropogenic greenhouse gas (GHG) emissions. In some instances, the value of rice fields as a supply of food has been increased by the addition of fish, particularly in

Asia (Xie *et al.*, 2011). Studies indicate, however, that stocking fish in rice plots may promote methane production (Frei and Becker, 2005), thus exacerbating GHG emissions.

Wetlands in Tanzania are extensively used for rice farming in combination with cattle grazing, and in certain parts of the country these agroecosystems contribute up to 98% of household food intake (McCartney and van Koppen, 2004; McCartney *et al.*, 2010). Many wetland agroecosystems provide multiple diverse options for meeting food security, especially for the people that are directly dependent upon them. Loss of these agroecosystems can have telling effects, not only on food supply, but also on the hydrological functions maintained by the wetlands.

Further, switching from one source of food to another within a wetland can have major implications for biodiversity, livelihoods and the distribution of benefits to people associated with one or the other activity, with both gains and losses, as shown by the case of Kolleru Lake in Andhra Pradesh, India and in the Testa, Brahmaputra and Padma river basins of Bangladesh (Nagabhatla *et al.*, 2012a,b; Senaratna Sellamuttu *et al.*, 2012). Starting in the early 1990s, the expansion of brackish water pond aquaculture in Thailand and Vietnam – at the expense of rice cultivation – has given rise to competing demands between both types of users, while causing dynamic changes in these wetland ecosystems (Szuster *et al.*, 2003; Dung *et al.*, 2009). In particular, the establishment of shrimp aquaculture has proven controversial; for example, in the coastal humid regions of South-east Asia and Latin America it has resulted in mangrove destruction on a large scale (Millennium Ecosystem Assessment, 2005). In places, shrimp culture is being developed further

inland to counter disease problems; this should reduce conflicts with mangroves, but may result in other negative environmental and social impacts. Elsewhere, integrated land-based marine aquaculture systems have been developed to optimize production, make input use more efficient and minimize waste discharges (Box 7.1).

Wetland Vulnerability and Implications for Food and Water Security

Wetland ecosystems are particularly vulnerable to changes in water quality and quantity (volume, flow pattern and timing), as these may damage their physical, chemical and biological properties (Gregory *et al.*, 2002; Alegria *et al.*, 2006; UNEP, 2006; Cho, 2007; Tran Huu *et al.*, 2009). Negative consequences for these ecosystems include river desiccation and functional fragmentation, groundwater depletion, water pollution and sedimentation, salinization and saltwater intrusion, soil erosion and nutrient depletion (Dugan *et al.*, 2007; Atapattu and Kodituwakku, 2009). Consequences such as these induce declines in biodiversity and other undesirable changes in the biota, e.g. trophic imbalance or simplification and loss of genetic populations (Dudgeon *et al.*, 2006). Problems relating to water imbalances in agroecosystems have dramatically changed the capacity of wetland ecosystems in the humid tropics to provide ecosystem services (Foley *et al.*, 2005).

Despite the importance of agriculture within wetlands, agriculture has been a major driver of wetland loss worldwide, both through water use and direct conversion. By 1985, an estimated 56–65% of inland and coastal marshes (including small lakes and ponds) had

---

Box 7.1. Horizontally integrated land-based marine wetland agroecosystems.

In Israel, tank-based culture systems have been developed combining, for instance: fish or abalone (edible sea snails) with seaweed; abalone, fish and seaweed; fish and shellfish; fish, microalgae and shellfish; fish, shellfish, abalone and seaweed. Constructed wetlands, planted with samphire (*Salicornia* spp.) that can be harvested for use as a vegetable, forage or biofuel have been evaluated to a limited extent for additional ecosystem services, including nutrient cycling (Bunting and Shpigel, 2009), but further work is required to assess the likely production from commercial-scale systems, the labour demands associated with management and harvesting, market perceptions and the risks associated with this strategy.

been drained for intensive agriculture in Europe and North America, 27% in Asia, 6% in South America and 2% in Africa (Millennium Ecosystem Assessment, 2005). Where historical records have permitted assessment, the rates of loss were shown to be high, for example, Valiela *et al.* (2001) found that more than one third of mangroves (35%) had been lost in the two decades up to the late 1990s, mainly to aquaculture (13.3% to shrimp farming and 4.9% to fish farming), deforestation (9.1%) and to upstream water diversions (3.9%). Throughout much of Asia, coastal ecosystems were extensively converted to agriculture during the 1960–1970s under the guise of what later became known as the Green Revolution. Operations in south-west Bangladesh and in West Bengal, India, and the associated costal engineering works, established large agricultural areas susceptible to secondary aquaculture development. Destructive practices such as this undermined both the processes that support ecosystems and the provision of associated services essential for human well-being (Millennium Ecosystem Assessment, 2005; Hoanh *et al.*, 2006; Molden, 2007; Atapattu and Kodituwakku, 2009).

Sub-Saharan Africa alone contains more than a million km^2 of wetlands, a large part of which are freshwater marshes and floodplains (Rebelo *et al.*, 2010). Out of more than 500,000 km^2 of wetlands designated as Ramsar sites, an estimated 93% support fisheries or agriculture, and 71% are facing threats due to these activities (Rebelo *et al.*, 2010). Indirectly, irrigation can threaten wetlands, not only by diverting fresh water, but also by reducing the capacity of rivers to transport nutrient-rich sediments that fertilize downstream wetlands and accrete to support the formation of new wetlands.

Excessive nutrient loading from fertilizers causes poor water quality and eutrophication of inland and coastal wetland systems (Lukatelich and McComb, 1986; Falconer, 2001; Molden, 2007; Deegan *et al.*, 2012). Chilka Lagoon in Odisha (formerly Orissa), India, for example, is affected by anthropogenic stresses as a result of agricultural practices and drainage in the catchment, which affect the water quality of the lagoon (Panigrahi *et al.*, 2007). Globally, in the coastal regions, agrochemical contamination is well documented to result in

bioaccumulation and have dire consequences on the numerous and diverse species that reside or feed in wetlands (Atapattu and Kodituwakku, 2009).

While contemplating impacts on wetlands caused by agriculture, we must also acknowledge the importance of wetlands in sustaining agriculture (both crop cultivation and livestock farming) and fisheries in developing countries, and the important role that wetland agriculture fulfills for livelihoods (Wood and van Halsema, 2008; McCartney *et al.*, 2010; Rebelo *et al.*, 2010). One way of doing this is by emphasizing multiple ecosystem services of agricultural wetlands and their value for livelihoods. In higher income countries, there is increasing realization of the magnitude, extent and importance of wetland services that have been lost; the consequences of which are often felt first among the farmers themselves. For instance, wetlands in the prairies of Canada have undergone a drastic conversion to agricultural land, but many farmers now realize that they suffer from decreased water availability as a result and are moving towards wetland restoration, as mentioned in Canada's fourth national report to the Convention on Biodiversity (CBD).

Urban wastewater is often discharged without adequate treatment and can negatively affect receiving water bodies. The productive use of wastewater to culture fish and irrigate rice and vegetables in the East Kolkata Wetlands of West Bengal, India, serves as an interesting example, however, of how urban wastewater has been turned into an asset (McInnes, 2010; Bunting *et al.*, 2011). Deliberate and planned use of wastewater for aquaculture was a feature of several large Asian cities, including Bangkok (Thailand), Hanoi and Ho Chi Minh (Vietnam) and Phnom Penh (Cambodia), but it has generally been phased out or lost owing to urban development (Little and Bunting, 2005; Bunting *et al.*, 2006). Contemporary use of wastewater for aquaculture continues widely, but is predominantly informal or unintentional, while responsible authorities may be reluctant to acknowledge that such practices occur. The cultivation of aquatic vegetables continues in peri-urban wetland agroecosystems around many cities in South-east Asia, but inorganic and chemical pollutants affecting wastewater quality constitute a risk to public and environmental health. Health risks can be reduced in a multi-barrier approach, but this would first require recognition of the use of wastewater in aquaculture (WHO, 2006).

Globally, wetlands are further threatened by human-induced climate change and the associated extreme weather events. Findings presented in the third and fourth assessment reports of the Intergovernmental Panel on Climate Change (IPCC) confirm that the changing water cycle is central to most of the climate change-related shifts in ecosystems and human well-being (Pachauri and Reisinger, 2007). By 2050, climate change is anticipated to have had significant impacts on coastal wetlands, both through changes in hydrological regimes and sea level rise. Future use of water and land for agriculture will further constrain the ability of wetland systems to respond and adapt to climate change. Coupled with ever-increasing human pressures, such as high-density populations and their associated needs, wetlands and their ecosystem services are seriously threatened unless the issues are urgently addressed and managed effectively. Hence, when water resource issues are to be addressed in climate change analyses and climate policy formulations, changes in the water cycle have to be considered as important starting points for interventions. Climate change variability will increase the need for improved water storage, and the role of wetlands and other water-based ecosystems in this, and the increased risk to wetlands of this adaptation strategy should be recognized (McCartney and Smakhtin, 2010). In view of the importance of wetlands in delivering ecosystem services, including the achievement of water and food security, the implication of most climate change scenarios is that it is more urgent than ever to achieve better management of wetland ecosystems in order to sustain water supplies and the other ecosystem services that they provide (Le Quesne et al., 2010).

Fisheries and Aquaculture in Wetland Agroecosystems

Fisheries and aquaculture are very important sources of food from wetland systems. Fishing techniques and aquaculture practices have been developed to exploit most wetland types (UNEP, 2010). Both fisheries and aquaculture provide synergies with rice cultivation (see above) by increasing water productivity as well as biodiversity. Variability and diversity within and among species and habitats are important for supporting this aquatic ecosystem service, and for increasing resilience (Molden, 2007). Culture-based fisheries, and stocking fish and other aquatic organisms in water bodies to grow for harvest with little further intervention, have been established mainly in seasonal wetlands, lakes and reservoirs, including water bodies in upland and highland areas of South and South-east Asia (Xie et al., 2011). Often developed to sustain livelihoods in fishing communities and enhance food security in poor and vulnerable rural communities, culture-based fisheries have also been proposed to increase employment and income from tourism and angling, or to enhance food production to alleviate fishing pressure on wild stocks (Lorenzen et al., 2012). Fish stocking and subsequent harvest has been proposed to facilitate the bio-manipulation of water bodies to enhance water quality characteristics, with the notable objectives of reducing invasive macrophyte communities (as well as harmful mosquito populations), increasing water clarity or sequestering nutrients. Stocking juvenile fish, however, constitutes a major cost, and there are ecological, social and economic risks associated with culture-based fisheries (Gurung, 2002).

Interventions such as stocking fish and other aquatic organisms in many wetlands have blurred the difference between capture fisheries, actions constituting fisheries enhancement and aquaculture. A systematic assessment of culture-based fisheries as an emerging aquatic resource management strategy has been undertaken by Lorenzen et al. (2012). According to Gurung (2002) carp have been stocked in several lakes in upland areas of Nepal to enhance production and reduce fishing pressure on 'thinly populated native species', while safeguarding employment and income for traditional fishing communities 'until measures for conservation practices of locally vulnerable species are developed'.

Inland capture fisheries landings, including fish, molluscs, crustaceans and other aquatic animals exceeded 11.2 million t in 2010, with the majority in Asia (68.7%), followed by Africa (22.9%), the Americas (4.9%), Europe (3.5%) and Oceania (0.1%) (FAO, 2012). Other assessments of small-scale fisheries in developing countries alone suggest that landings are even more significant in those countries, with an estimated 14 million t caught annually (Mills *et al.*, 2010). These catches provide food and livelihoods for 60.4 million people, 33 million of whom are women (UNEP, 2010). A wide range of aquatic ecosystems are important for fisheries, perhaps the most obvious being those where fish and other aquatic animals are caught. Breeding and nursery sites, which may be quite distant from fishing areas, also play a critical role in the life cycles of exploited stocks, and these could be managed better in the wider landscape of agroecosystems (Dugan *et al.*, 2007). Similarly, terrestrial ecosystems and catchment land use practices influence the hydrology and quality characteristics of water resources, which, in turn, are critically important in governing the types of species that can survive in certain habitats (Welcomme *et al.*, 2010). Stocking aquatic animals in predominantly wetland agroecosystems, with interconnected field and pond systems, may make a significant contribution to food security and nutrition in farming households and local communities (Xie *et al.*, 2011). Appropriate management and governance arrangements are required to ensure that costs and benefits are distributed equitably, and that any proposed changes in access arrangements consider the needs of poor and landless groups (FAO, 2010).

Aquaculture development and fisheries depend on the appropriation of various environmental services from aquatic ecosystems; these include clean and oxygenated water for physical support and respiration, inputs of seed, feed and detritus, waste removal, nutrient assimilation and carbon sequestration (Beveridge *et al.*, 1997). The failure of many apparently promising aquaculture ventures has occurred when the capacity of ecosystems to meet the cumulative demand for environmental goods and services from rapidly growing numbers of farms and culture units has been exceeded (Bostock *et al.*, 2010). An example is the proliferation of cage-based aquaculture in the Saguling Reservoir, Indonesia, where self-pollution was implicated in causing massive fish kills (Hart *et al.*, 2002). Early assessments of the appropriation of environmental goods and services by aquaculture systems intimated that the ecological footprints (expressed as m^2 supporting ecosystem/m^2 culture facility),[2] were larger for more intensive production systems (Berg *et al.*, 1996; Folke *et al.*, 1998). Subsequent reassessment, however, showed that some goods and services were used more efficiently in the intensive production systems than in semi-intensive systems (Bunting, 2001). While expressing ecological footprints per unit area of production system helps to visualize the dependence on the ecosystem support area, assessment per unit of production permits a more rational appraisal of alternative management strategies for the same culture area.

Constructing big dams and the extensive development of small-to-medium sized structures for hydroelectric power has had widespread negative ecological and social impacts. Notable ecological impacts include: immediate devastation wrought on inundated aquatic ecosystems; impacts on downstream wetlands and wetland agroecosystems; and disruption to connectivity and environmental flows between ecosystems. Dam construction may result in fertile land used for cereal crop production being inundated, so threatening food security; even when higher value products can be caught from new water bodies or extracted from forests, the equilibrium of survival may mean that people are unable to buy sufficient staple foods to meet their needs. Added pressure on forest resources affects catchment dynamics, and the lure of valuable harvests may attract migrant fishers with the skills and technology to catch fish in deep lakes, and consequently result in potential benefits not reaching local communities (Nguyen Thi *et al.*, 2010).

Water management in humid agroecosystems often involves multiple uses of water and can be further enhanced by considering the whole range of ecosystem services through a gender-sensitive approach. Some good examples are the integration of aquaculture into various agroecosystems, such as livestock–aquaculture integration, rice–fish culture, aquaculture in irrigation reservoirs and water management schemes, and wastewater-fed aquaculture. Evaluation of the full range of provisioning ecosystem services from aquatic ecosystems, not only fish, is vital if the true value of wetlands and wetland agroecosystems in the livelihoods of men and women, and in local and national economies, is to be accounted for and safeguarded.

The current appropriation of aquatic ecosystem services is often not sustainable; this is the case with fishing in most waterways and wetlands, and with the majority of semi-intensive and intensive aquaculture production around the world. As with the assessment of marine capture fisheries, there must be concern over introducing shifting baselines (Pauly, 1995), and setting overly generous limits or inappropriate conservation goals. It is critical to maintain a balance between fisheries – often the most obvious benefit derived from aquatic ecosystems – and the continued provision of stocks and flows of other ecosystem services, as these may actually benefit more people and make a more significant contribution to the well-being and resilience of poor women and men, marginal groups, local communities or regional populations (Welcomme *et al.*, 2010). Moreover, assessment and allocation of water resources must also account for environmental water requirements (Gichuki *et al.*, 2009).

Integration of Aquaculture in Agroecosystems

Livestock, agriculture, horticulture, aquaculture and fisheries production have been closely integrated in iconic farming systems for hundreds of years. Examples include: dyke pond farming in the Pearl River Delta in Guangdong Province and rice–fish culture in Zhejiang Province, China; canal dyke culture in Thailand and Vietnam; *chinampa* cultivation (growing crops on artificial islands in shallow lake beds) in Mexico; and taro cultivation with fish ponds in Hawaiian *apupua'a* agro-ecosystems (an *apupua'a* is a designated subdivision of a Hawaiian island) (Beveridge and Little, 2002). Several of these traditional systems have virtually disappeared and most

are now under immense pressure to change, owing to greater concentration on high-value, cash crop production supported by external technology (formulated feeds, inorganic fertilizers, agrochemicals, mechanical pumps, aerators and filters, and agricultural machinery). Such intensification of production is often precipitated by the need to increase economic returns from land holdings that have significantly appreciated in value over recent years.

Globalization and the expansion of international trade are major driving forces that are exerting pressure to convert natural wetlands and intensify production in wetland agroecosystems (see Chapter 2). Consequently, trade-based mechanisms such as product certification and ecolabelling might be considered to counter such forces. Fundamental reform may be required, however, to shift aquaculture towards a more sustainable development pathway. Authorities should remove subsidies for unsustainable practices, force producers to account for negative environmental costs and promote the adoption of better management practices. Semi-intensive pond-based fish production that depends on organic and, increasingly, inorganic fertilizer to stimulate the natural production of food to supplement low-cost feeds with modest protein contents remains widespread in China and throughout much of Asia. Prevailing market forces could conceivably compel producers to opt for intensive production that would be totally dependent on high-protein formulated feeds. This would result in the loss of ecosystem services associated with semi-intensive production, notably the managed disposal of large volumes of organic waste, including manure, and agricultural and food processing by-products.

Promising approaches to productive multiple use of water resources that persist include rice–fish farming and the integration of aquaculture and culture-based fisheries in reservoirs, and these are discussed further below. Negative environmental externalities associated with intensive farming become more apparent as the full cost of external feed, fertilizer, fuel and technology inputs are accounted for in cost–benefit or life-cycle assessments (Hall *et al.*, 2011). Together, these are likely to influence policy making and consumer attitudes, and may signal a renascence for traditional resource-efficient and conserving farming systems. Therefore, it is important to preserve knowledge and, ideally, examples of such integrated systems to guide and inform emerging ecocultures. Conditions, constraints and water use efficiencies in various aquaculture-based agro-ecosystem combinations are summarized in Table 7.2.

Table 7.2. Integration of aquaculture practices with other activities to optimize efficiency and increase water productivity (adapted from Bunting, 2013).

Management practices	Constraints and conditions	Potential water use efficiency outcomes
Livestock–aquaculture		
• Ducks and geese foraging on ponds	• Possible pathogen and disease transfers within integrated systems	• Multiple products from ponds and lakes with lower water footprints
• Wildfowl and poultry housed over fish ponds	• Chemical treatments and dietary supplements for livestock may affect production and accumulate in aquaculture components	• Enhanced environmental protection of receiving water through better on-farm waste management and nutrient recycling
• Waste from pigs and cattle directed to fish ponds for treatment and nutrient recycling		
• Plant and fish biomass cultivated using solid and liquid waste fed to livestock	• Excessive waste loadings or perturbations affecting the ecological balance of the pond can result in low oxygen levels and fish health problems and mortality	• Aquaculture of biomass and fodder crops helps to avoid public health risks and consumer acceptance of aquatic products grown using waste resources
Aquaculture in irrigation and water management schemes		
• Fish cages in irrigation channels in India and Sri Lanka	• Excessive flow rates can have an impact on animal welfare and make food unavailable	• Nature of aquaculture means water is conserved, potentially with higher nutrient content, thus enhancing crop production
• Culture-based fisheries in domestic supply and irrigation reservoirs	• Debris can block mesh, reducing flow rates and causing physical damage to fish cages	• Aquatic species may predate upon disease vectors, crop pests and weeds
• Aquaculture in traditional irrigation structures within microcatchments in Sri Lanka	• Management must balance irrigation and aquaculture demands	• Integration of aquaculture activities may enhance nutrient cycling and uptake by plants under irrigation
• Fish culture in irrigated rice fields and farmer-managed systems in Africa and Asia	• New structures may be needed to sustain fish populations during low water periods	
	• Agrochemicals in extended irrigation systems and adjacent areas can affect aquaculture productivity and may constitute a public health concern	

Management practices	Constraints and conditions	Potential water use efficiency outcomes
Aquaculture in water storage reservoirs		
• Fish cages in reservoirs for hydroelectric power generation • Culture-based fisheries in water storage and hydroelectric reservoirs • Polyculture in urban and peri-urban water bodies, primarily for floodwater discharge and amenity	• Inappropriate reservoir bed preparation, presence of submerged structures (including downed trees) and routine drop down may reduce the area suited to aquaculture development • Rapid drop down may damage physical cage structures • Changes in access and use rights associated with aquaculture development may cause social problems	• Multiple use of water in reservoirs could contribute to increased revenue generation and alternative livelihoods for displaced or marginal communities • Appropriate species selection for aquaculture could contribute to weed control and enhance water quality in reservoirs
Aquaculture in saline drainage and wastewater		
• Aquaculture in saline groundwater evaporation basins in Australia • Fish culture in saline wastewater from industrial processes and desalinization	• Variation in salinity levels and possible extremes may constrain species selection or culture duration • Low production rates as compared with prevailing commercial operations suggest need for further assessment of financial and economic attributes	• Exploitation of saline water resources through integration of aquaculture can contribute to overall farm productivity and generate new income streams • Economic benefits of integrating aquaculture, salt-tolerant crop production and salt harvesting could help offset costs of controlling saline groundwater problems
Aquaculture in thermal effluents and cooling water		
• Production of juvenile fish in cooling water effluents from nuclear power stations in France • Farming marine worms in thermal effluents in the UK	• Chemicals used to clean power stations and variations in water temperature may affect growth and product quality • Farming species for human consumption may pose unacceptable health risks or not gain consumer acceptance	• Retention of thermal effluents for aquaculture production can facilitate heat dissipation and contribute to meeting statutory discharge standards • Exploitation of thermal effluents can help to avoid greenhouse gas emissions associated with heating water for culturing cold-intolerant species

Continued

Table 7.2. Continued

Management practices	Constraints and conditions	Potential water use efficiency outcomes
Urban and peri-urban aquaculture		
• Fish cages in canals and lakes in Bangladesh and Vietnam • Fish culture in canals, lakes, ponds and borrow pits in peri-urban areas throughout Asia • Macrophyte cultivation in drainage canals and low-lying water bodies, e.g. Bangkok (Thailand), Hanoi (Vietnam), Phnom Penh (Cambodia) • Aquaculture exploiting food and drink production and processing by-products	• Multiple use of urban and peri-urban water bodies may mean hydrology is out of the control of aquaculture producers and associated operational constraints result in suboptimal management • Risks from pollution and poaching may constrain aquaculture development • Insecure land tenure and pressure from urban residential and industrial development may constrain investment in aquaculture systems	• Floodwater storage and groundwater recharge associated with extensive wastewater-fed aquaculture operations can contribute to stabilizing local hydrological conditions • Vigilance of aquaculture producers helps in monitoring pollution and safeguarding water quality for other users
Aquaculture in multi-purpose household ponds		
• Fish culture in small ponds used primarily for domestic and agricultural purposes • Composite fish culture in rainwater harvesting structures	• Introduction of aquaculture can cause conflicts with other agricultural and domestic uses of household ponds • Inclusion of aquaculture in rainwater harvesting ponds may constrain the use of water use for other crops and incur financial risks	• Appropriate integration of aquaculture into household ponds can contribute to food security and livelihood outcomes without reducing water availability for other purposes • Aquaculture in ponds can help reduce pressure on the provisioning ecosystems services of natural water bodies
Wastewater-fed aquaculture		
• Intentional use of wastewater to supply water and nutrients for aquaculture • Lagoon-based sewage treatment systems incorporating fish ponds developed under the Ganges Action Plan initiative, India • Fish culture in 3900 ha of ponds in the East Kolkata Wetlands, West Bengal, India • Duckweed cultivation on wastewater in the UK for processing to biofuel	• Health risks posed by waste-water use for aquaculture demand that appropriate treatment and control measures are adopted • Consumer perceptions, prevailing beliefs and institutional barriers may constrain development • Land area required for combined wastewater treatment and reuse through aquaculture may prohibit development	• Management of wastewater promoted by integration of aquaculture can help operators meet statutory discharge standards and help safeguard public health • Wastewater reuse through aquaculture can help protect the quality of water bodies receiving discharge from the system • Exploitation of wastewater flows for biomass production could help alleviate pressure on freshwater resources

Aquaculture in rice fields

A special case of integrated aquaculture that has a long tradition is fish culture in rice fields. In the discussion of wetlands it is also important to recognize the synergies between fisheries and rice cultivation as practised in South-east Asia and elsewhere. These practices may create agroecosystems that have higher biodiversity and increased water productivity, although there are examples where biodiversity declines (Bellio *et al.*, 2009). For example, the conversion of traditional deep water rice cultivation on floodplains in Asia to irrigated systems planted with high-yielding varieties has been implicated in the loss of both aquatic biodiversity and indigenous rice varieties. Culturing fish in rice fields can help to control pests and weeds, promote nutrient availability to rice plants and increase nutritional benefits and financial returns from what are widely regarded as low input, environmentally friendly and more sustainable farming systems. Integrating fish culture into irrigated and rain-fed rice fields also makes more effective use of appropriated water resources.

The culture of fish in rice fields has been traditionally practised in China, Japan and Java (Indonesia); more recently, rice–fish culture has been introduced by development agencies and extension services to many countries in Asia and to a growing number in Africa. However, integrated culture of rice and fish requires refined farm management approaches to coordinate rice production and fish culture practices, with increased dependency on reliable water supplies. Often, lack of this expertise, combined with poor market linkages (unreliable fish seed production, and poor infrastructure for distribution of harvested fish) has constrained widespread and long-lasting adoption. Where rice–fish culture has been adopted widely, e.g. in north-east Thailand and West Java, it has made an important contribution to incomes and food security in poor and marginal farming communities. Perceived declines in the availability of wild fish and well-developed trading networks for fish seed from private hatcheries have stimulated the adoption of rice–fish culture in north-east Thailand. Paddy fields can also be used as nurseries for fingerlings, and these can be sold to stock ponds; such strategies have great potential in facilitating the decentralization of fish seed supply and promoting aquaculture development.

Low-input rice–fish culture could be a viable alternative, measured in conventional financial terms and based on standard risk assessment criteria, as farmers face increasing bills for fertilizers and pesticides to maintain yields in high-input, irrigated, monoculture rice production. Farmers should be supported in assessing their prospects for adopting rice–fish culture and, where demand exists, action should be taken to ensure a functional enabling institutional environment. The successful development of rice–fish culture has been attributed to: the adaptation of traditional water management approaches to accommodate fish culture; appropriate extension services, training and capacity building; and access to quality fish seed of the appropriate species.

Aquaculture in irrigation systems

New capture fisheries are often cited as a secondary benefit associated with reservoirs developed for irrigation purposes, but the timely colonization by species suited to reservoir conditions and valued by fishermen is not guaranteed. Furthermore, unrestricted and unregulated fishing could limit the establishment of a substantial, self-reproducing stock of desirable species (Munro *et al.*, 1990). Consequently, establishing a culture-based fishery, or fish culture in pens or cages, may be proposed as alternative solutions (Lorenzen *et al.*, 2012). The infrastructure to support culture-based fisheries, including hatcheries, is often commissioned as part of reservoir construction projects, with fishing rights being leased out to local groups. The construction of pens and cages can be used to partition the available water resource, potentially enabling displaced or landless peoples to gain some form of employment and security; then again, the costs of constructing and stocking such structures can be prohibitive, often leading to rich individuals and commercial enterprises dominating the available resources (Beveridge, 2004). Smaller cages can be deployed in irrigation canals, but flow rates and regimes must be suitable and the requirements of cage operators must be considered in the overall planning and management of the irrigation system.

Rapid uncontrollable expansion of aquaculture in larger irrigation reservoirs can result in access to fishing grounds and navigational routes being disrupted and this, in turn, can lead to social tension and an inequitable flow of benefits that tends to be of advantage to the more affluent sections of society (Beveridge and Phillips, 1993). Drawdown and the presence of submerged trees can restrict the area available for the development of cage-based aquaculture (Table 7.2). Rapid drawdown can cause physical damage to cages and lead to upwelling of deoxygenated water from the hypolimnion, which can cause mortalities in overlying fish cages (e.g. Lake Sampaloc, Philippines). Fluctuating water levels can be a serious problem for both fish production and fisheries, as well as for cage aquaculture, in reservoirs used for irrigation and hydroelectric power generation. The uncontrolled development of aquaculture and associated waste discharges can lead to deterioration of water quality, reducing ecosystem services (drinking water, fish) and posing problems for downstream water users (Beveridge and Stewart, 1998). A strong environmental policy and appropriate governance arrangements, including the implementation of adaptive management, are needed to ensure that cage development delivers the anticipated economic, social and food security benefits.

Wetlands Assessment and Management

Wetlands contain biodiversity of exceptional conservation significance and support many unique ecosystems and a wide array of globally threatened species. At the same time, they typically form an essential component of local, national and even regional economies, as well as underpinning the livelihoods of many rural communities. Yet, despite their importance, they are under increasing pressure. Often, wetlands and wetland agroecosystems have been managed in isolation – disconnected physically and in policy making and planning

from the associated river basin system. Weak consideration of wetlands in decision making remains one of the major factors leading to their degradation (Horwitz and Finlayson, 2011). Management decisions affecting wetlands rarely consider the wider biological, ecological, developmental or economic values of wetlands, as they are challenging and costly to assess.

Various agricultural practices can be advocated that promote the wise use of wetland ecosystems while ensuring sustainable development. Ecosystem analysis must be integrated with assessments of associated livelihood strategies if resulting management plans are to gain broad-based support and address the underlying pressures and unsustainable use practices. Knowledge of the interconnectedness of wetlands and fisheries provides valuable examples in this regard (e.g. Smith *et al.*, 2005).

The adoption of strategies (i.e. of the relevant provisions of the CBD and Ramsar Convention) that work towards the environmental management of these ecosystems would link environmental stewardship directly to poverty alleviation, food security and the quality of water in wetlands (Millennium Ecosystem Assessment, 2005). Food production practices and wetland management plans should be jointly assessed by concerned stakeholder groups, and measures taken to ensure that the demands placed on the environment are within acceptable limits (see Box 7.2). Such assessments should be conducted with groups that have been disaggregated on the basis of wealth, gender and generation to account for differences in needs and priorities, otherwise the outcomes risk further disadvantaging poor and vulnerable groups and, ultimately, undermining conservation and management initiatives (see Box 7.3). Alternative approaches include a combination of limited data collection and modelling – instead of full-scale assessments – to develop options for wetland management (Cools *et al.*, 2012, 2013; Johnston *et al.*, 2013).

If better management is sought, the development, assessment and diffusion of applicable technologies that increase the production of food per unit of water, without

---

Box 7.2. Integrated wetland assessment in Cambodia and Tanzania.

The International Union for Conservation of Nature (IUCN) has developed a toolkit of methodologies for assessing the value of wetland biodiversity to livelihoods, particularly those of the poorest, and finding ways to clearly present this information to decision makers (Springate-Baginski *et al.*, 2009). The methodologies are integrated and incorporate biodiversity, economics and livelihoods approaches. The toolkit was put into practice in two demonstration sites: the Stung Treng Ramsar site in Cambodia and Mtanza-Msona village in Tanzania (Allen and Springate-Baginski, 2008). Following initial preparation and orientation activities, notably clarifying the management objectives of stakeholders, the integrated wetland assessment fieldwork was completed, and integrated reports on the livelihood, biodiversity and economic values of the areas were prepared.

These assessments yielded detailed scientific and management information, including GIS (geographical information system) maps and databases, which document key values and overlaps between threatened species and areas of high human dependence. Information obtained in the Stung Treng Ramsar site was included in the management and zoning plan for this site, towards supporting pro-poor wetland conservation and sustainable use to the benefit of local livelihoods and biodiversity. Data obtained from the second demonstration site helped local communities to understand the importance of wetland resources in their livelihoods.

The main output of the project was *An Integrated Wetland Assessment Toolkit: A Guide to Good Practice* (Springate-Baginski *et al.*, 2009). This guide provides a set of integrated assessment methods that combine and investigate the links between biodiversity, economics and livelihoods, with a particular focus on strengthening pro-poor approaches to wetland management. It aims to assist in overcoming the current methodological and information gaps in wetland planning, to factor wetland values into conservation and development decision making and management planning, and to assist in identifying areas of potential conflicting priorities. The toolkit is expected to be of use to wetland site managers, conservation and development planners, and researchers from both natural and social science disciplines.

The studies in Cambodia and Tanzania brought experts from social, ecological and economic backgrounds to work together. It was not easy to convince them of the value of the work in each of the other two disciplines. For example, it was challenging, but ultimately successful, to convince the social scientists of the value of biodiversity assessment; vice versa, it was challenging to find good models and tools as examples of integrated work (Allen and Springate-Baginski, 2008).

---

harmful trade-offs, is both feasible and essential (see Chapter 8). Though such technologies have already been identified and are available, the majority of countries have failed to promote them and to penalize more damaging practices, and less developed countries lack the financial resources to improve their capacity to adopt such approaches (Millennium Ecosystem Assessment, 2005). However, certain strategies can be adopted in order to realign policies on agriculture and wetlands (Peden *et al.*, 2005; Molden, 2007; Wood and van Halsema, 2008; McCartney *et al.*, 2010):

- Improve the agricultural practices of farmers in ways that positively influence wetlands, while at the same time not compromising livelihoods. This can be done by: increasing agricultural productivity (intensification) without expanding land area or water use, thereby not com-

promising the water regulatory functions of wetlands; shifting from irrigation to rainfed agriculture; and improving soil management.

- Adopt supporting strategies that maintain and improve wetland ecosystem services so that a broader range of stakeholders, including the rural poor, receive the benefits.
- Assess water use by the surrounding agroecosystems and adapt its use to be in harmony with a sustainable supply, using trade-off analyses.
- Improve land and water management techniques after a comprehensive evaluation of the social and ecological products and services supported by the wetlands for women and men.
- Provide alternate livestock drinking sites away from sensitive wetland areas, not only for the benefit of the wetlands, but also as a means to reduce animal health risks.

Box 7.3. Wetlands and livelihoods in South Africa (WWF, 2009)

The Sand River's upper catchment wetlands in South Africa's Limpopo Province are within densely populated communal lands. The wetland farmers, 90% of whom are women, are among the poorest of the country and depend on these freshwater ecosystems as their only source of food. However, their farming practices, passed from generation to generation, are causing increased erosion, increased desiccation, poor soil fertility and low productivity. In partnership with the Association for Water and Rural Development (AWARD), the World Wildlife Fund for Nature (WWF) South Africa Programme Office started a project to recover the ecological functions of the Sand River's wetlands while improving the livelihoods of the communities living in this area. The project aims to promote awareness of the value of goods and services of the Sand River wetlands in providing livelihood security to poor rural communities, and to develop good agricultural practices among wetland farmers and harvesters in the Sand River Basin.

The project started by evaluating the nature and intensity of farming practices in the wetlands; detailed and rapid appraisals on 60 plots were completed using interviews, field assessments and documentary photographs. The appraisals confirmed erosion, desiccation and poor soil fertility as the main negative outcomes from farming practices. Because wetland farmers relied very much on the wetlands for their livelihoods, it was assumed that they understood their value, but this was shown not to be true, and getting farmers to change their practices and think about long-term management of the wetlands was a challenge. Based on this information, all 60 farmers were grouped according to shared issues, and were engaged in a series of workshops and field visits, whereby they were introduced to basic wetland concepts, conservation tillage methods and good wetland practices. During these workshops, discussions about the need for change were carried out so that farmers could understand the connection between their livelihoods and long-term wetland security and functioning. Farmers then designed their own action plans as well as impact indicators.

These actions were implemented and their impact upon agricultural practices and the state of the wetlands was determined using the indicators that had been defined. An obstacle to this was the inadequate communication and lack of self-organization among farmers. Poor trust hampered exchange of knowledge about the actions implemented, although with the support of the project team, in time the farmers understood the importance of working together to find ways to use the wetlands more sustainably. They also became aware that a number of the problems they faced had their origins at the microcatchment level, and that working with other stakeholders was needed. Hence, they started working on reducing livestock, avoiding damaging crops, preventing gully erosion and managing the large quantities of water entering the wetland from the surrounding villages.

- Improve awareness among all stakeholders who are involved in agricultural water management, and improve their understanding of ecosystem services.
- Improve the inventories, assessment and monitoring of interactions with agroecosystem changes, and of changes to the surrounding wetland. Apply environmental monitoring and decision support systems that involve the affected local communities.
- For each water use activity, identify who are the winners and losers among men and women and affected social groups, and determine the costs and benefits incurred by each and look for ways to transfer costs into incentives to farm more sustainably.
- Adopt an integrated approach to water management that considers the whole catchment, its land use and the water and

wetland ecosystems within it, in a way that balances the multiple water requirements for livelihoods along with the needs of the different ecological processes of wetland ecosystem services.

When planning and implementing stocking strategies, appropriate risk assessments and control measures should be employed to protect native fish populations and ensure that other species are not negatively affected (Lorenzen et al., 2012). The potential social, cultural and environmental impacts of such interventions demand careful assessment prior to their implementation. It is important to adopt a gendered approach in developing such strategies, as there are gender-related differences both in resource access and use, and in the accrual of productive benefits and their distribution among family members (e.g. World Bank *et al.*, 2009).

Greater understanding is required of the continuum of practices from capture fisheries via stock enhancement to fully-fledged aquaculture, if natural resource managers and responsible authorities are to account for such activities in planning and policy making. Notably, it needs to be understood that management regimes may shift as a result of perceived production risks, environmental change, emerging market opportunities, and evolving governance structures and organizational arrangements.

As an alternative to stocking natural water bodies and in response to environmental concerns over intensive, monoculture-based aquaculture, horizontally integrated or land-based integrated multi-trophic aquaculture (IMTA) systems have been developed (see Box 7.1). Within such systems, farmers use formulated feeds for some species, e.g. fish, shrimp or abalone, but these are cultured together with other organisms, notably microalgae, shellfish and seaweed, that convert nutrients released from the fed component to harvestable biomass. This can be used as a supplementary food source, hence reducing the demand for formulated feed, or generating additional revenue, thereby increasing the efficiency and productivity of the system. Also, the integration of aquaculture with other activities can enhance water use efficiency and productivity, though the opportunities and constraints associated with different strategies vary (Table 7.2), and the attendant risks and potential benefits may be difficult to quantify (Bunting and Shpigel, 2009; Troell *et al.*, 2009).

Integrated systems permit the generation of higher revenues and more regular cash flows from water pumped ashore or from underground, or available via tidal exchange. A pond-based system combining fish, microalgae and shellfish developed on the Atlantic coast of France received water from a tidally filled reservoir. The disadvantage of the system was that reservoir capacity limited the biomass of fish cultured and, consequently, the amount of integrated production that could be maintained in the system. In tropical coastal areas, integrated farming systems combining pond-based fish and shrimp production with shellfish and seaweed production have been developed, although high concentrations of suspended solids can constrain shellfish growth, and high turbidity and grazing can limit algal production. In such systems, mangrove stands have been used to condition incoming water and treat aquaculture wastewater.

The economies of integration that are associated with horizontally integrated systems, using the same water, feed inputs, infrastructure, equipment and labour to produce multiple crops, appear to offer a potential advantage over monoculture systems as they provide a wider range of ecosystem services. Opportunities to develop comparable systems in freshwater settings could be explored, as well as assessments made to determine their impact on stocks and flows of ecosystem services within and outside the system. However, integration places new demands on farmers in terms of skills and knowledge requirements, and results in additional risks, in particular related to engineering requirements and pests and diseases; it also poses new and poorly defined statutory and marketing challenges.

Conclusions

Wetlands across the world play a critical role in the provision of freshwater for human consumption and agriculture, while both fresh and saline waters provide food security by supporting fisheries, aquaculture and other related activities. Many wetland agroecosystems represent multiple and diverse options for meeting food security, as well as for meeting basic human needs for water, especially for populations that are directly dependent upon them. Where wetlands themselves are used for agricultural production, as in many parts of Africa and Asia, they help to safeguard the livelihoods of the rural poor; but the increasing deterioration in the condition and, thus, the resilience to future shocks of these natural systems is increasingly putting in jeopardy such safeguards for people.

Urgent steps are needed to protect biodiversity-rich wetland ecosystems, with their multitude of functions and services, as well as the livelihoods and well-being of the dependent communities. Once these areas are identified as wetland agroecosystems with their own set of ecosystem services, effective water management can be put in place with the minimum of trade-offs against other services. The monitoring of wetland functions and services is crucial to ensure the continuation of wetland ecosystems and safeguard their role in secure, high-quality food and water provision, as well as in many other critical and related ecosystem services, including flood protection and climate regulation. A number of priorities can be identified for action, investment and policy to conserve wetlands and promote sustainable development:

- An ecosystem-based approach to wetland management should be adopted that takes into account the contribution of such areas to the livelihoods of primary stakeholders, notably poor and marginal groups, resource users and local communities.
- Investment is needed to identify and promote approaches to food production supported by wetlands that are sustainable and appropriate, given the local social–ecological conditions and food security needs.
- Formulation of wetland management plans should be based on principles of integrated wetland assessment that enable combined biodiversity, economic and livelihoods analysis across different disciplines and sectors.
- Wetlands sustain an array of ecosystem services contributing to human well-being

and food security, but polices and management plans are needed to ensure that environmental stocks and flows are protected so as to safeguard aquatic biodiversity.

Notes

[1] Wetlands were defined under the Ramsar Convention as 'areas of marsh, fen, peatland or water, whether natural or artificial, permanent or temporary, with water that is static or flowing, fresh, brackish or salt, including areas of marine water the depth of which at low tide does not exceed six metres' (Ramsar Convention Secretariat, 2011).

[2] Assessments of ecological footprints have the potential to highlight disparities between the demand and supply of ecosystem services for particular culture systems but care is needed in the calculation and interpretation of footprints, especially with respect to geographical and temporal differences in the location and availability of goods and services. Appropriation of goods and services by other sectors also needs to be considered and environmental stocks and flows maintained. Approaches to supplement ecological goods and services in certain cases have been proposed but it is difficult to replicate natural processes in ecologically engineered systems. Moreover, the development of such systems may cause further environmental and financial impacts and, being ecologically-based, operation and performance of such systems will be highly influenced by prevailing environmental conditions, notably temperature and light levels, and vulnerable to other natural occurrences such as storms, pests and diseases.

References

Alegria, H., Bidleman, T.F. and Figueroa, M.S. (2006) Organochlorine pesticides in the ambient air of Chiapas, Mexico. *Environmental Pollution* 140, 483–491. doi:10.1016/j.envpol.2005.08.007

Allen, D. and Springate-Baginski, O. (2008) Good practice toolkit for integrated wetland assessment integrating species assessment, livelihoods assessment and ecosystem service valuation for improved wetland planning and management. Presentation from the Water Pavilion at the: *IUCN World Conservation Congress, Barcelona 5–14 October 2008.*

Amilhat, E., Lorenzen, K., Morales, E.J., Yakupitiyage, A. and Little, D.C. (2009) Fisheries production in Southeast Asian farmer managed aquatic system (FMAS) I. Characterisation of systems. *Aquaculture* 296, 219–226.

Atapattu, S.S. and Kodituwakku, D.C. (2009) Agriculture in South Asia and its implications on downstream health and sustainability: a review. *Agricultural Water Management* 96, 361–373. doi:10.1016/j.agwat.2008.09.028

Bellio, M.G., Kingsford, R.T. and Kotogama, S.W. (2009) Natural versus artificial – wetlands and their waterbirds in Sri Lanka. *Biological Conservation* 142, 3076–3085.

Berg, H., Michelsen, P., Troell, M., Folke, C. and Kautsky, N. (1996) Managing aquaculture for sustainability in tropical Lake Kariba, Zimbabwe. *Ecological Economics* 18, 141–159. doi:10.1016/0921-8009(96)00018–3

Beveridge, M.C.M. (2004) *Cage Aquaculture*, 3rd edn. Blackwell, Oxford, UK.

Beveridge, M.C.M. and Little, D.C. (2002) The history of aquaculture in traditional societies. In: Costa-Pierce, B.A. (ed.) *Ecological Aquaculture. The Evolution of the Blue Revolution.* Blackwell, Oxford, UK, pp. 3–29.

Beveridge, M.C.M. and Phillips, M.J. (1993) Environmental impact of tropical inland aquaculture. In: Pullin, R.S.V, Rosenthal, H. and Maclean, J.L. (eds) *Environment and Aquaculture in Developing Countries, Bellagio, Italy, 17–22 September 1990.* ICLARM Conference Proceedings 31, International Center for Living Aquatic Resources Management [now WorldFish], Manila and Deutsche Gesellschaft für Technische Zusammenarbeit Frankfurt/Main, [now Deutsche Gesellschaft für Internationale Zusammenarbeit, Bonn], Germany, pp. 213–236.

Beveridge M.C.M. and Stewart J.A. (1998) Cage culture: limitations in lakes and reservoirs. In: Petr, T. (ed.) *Inland Fishery Enhancements.* FAO Technical Paper 374, Food and Agriculture Organization of the United Nations (FAO), Rome, pp. 263–279.

Beveridge, M.C.M., Phillips, M.J. and Macintosh, D.J. (1997) Aquaculture and the environment: the supply and demand for environmental goods and services by Asian aquaculture and the implications for sustainability. *Aquaculture Research* 28, 797–807. doi:10.1046/j.1365-2109.1997.00944.x

Bostock, J., McAndrew, B., Richards, R., Jauncey, K., Telfer, T., Lorenzen, K., Little, D., Ross, L., Handisyde, N., Gatward, I. and Corner R. (2010) Aquaculture: global status and trends. *Philosophical Transactions of the Royal Society B* 365, 2897–2912. doi:10.1098/rstb.2010.0170

Bullock, A. and Acreman, M. (2003) The role of wetlands in the hydrological cycle. *Hydrology and Earth System Sciences* 7, 358–389.

Bunting, S.W. (2001) Appropriation of environmental goods and services by aquaculture: a re-assessment employing the ecological footprint methodology and implications for horizontal integration. *Aquaculture Research* 32, 605–609. doi:10.1046/j.1365-2109.2001.00563.x

Bunting, S.W. (2013) *Principles of Sustainable Aquaculture: Promoting Social, Economic and Environmental Resilience.* Earthscan, London, from Routledge, Oxford, UK.

Bunting, S.W. and Shpigel, M. (2009) Evaluating the economic potential of horizontally integrated land-based marine aquaculture. *Aquaculture* 294, 43–51. doi:10.1016/j.aquaculture.2009.04.017

Bunting, S., Little, D. and Leschen, W. (2006) Urban aquatic production. In: van Veenhuizen, R. (ed.) *Cities Farming for the Future: Urban Agriculture for Green and Productive Cities.* Published for RUAF Foundation (Resource Centres on Urban Agriculture and Food Security), International Institute of Rural Reconstruction (IIRR) and International Development Research Centre (IDRC) by IIRR, Silang, Cavite, Philippines and ETC Urban Agriculture, Leusden, the Netherlands, pp. 379–410 (unnumbered). Available at: http://www.ruaf.org/sites/default/files/Chapter%2013.pdf (accessed February 2013).

Bunting, S.W., Edwards, P. and Kundu, N. (2011) *Environmental Management Manual: East Kolkata Wetlands.* MANAK Publications, New Delhi.

Cho, D.-O. (2007) The evolution and resolution of conflicts on Saemangeum Reclamation Project. *Ocean and Coastal Management* 50, 930–944. doi:10.1016/j.ocecoaman.2007.02.005

Cools, J., Diallo, M., Boelee, E., Liersch, S., Coertjens, D., Vandenberghe, V. and Kone, B. (2012) Integrating human health into wetland management for the Inner Niger Delta, Mali. *Environmental Science and Policy* (in press). doi:10.1016/j.envsci.2012.09.011

Cools, J., Johnston, R., Hattermann, F.F., Douven, W. and Zsuffa, I. (2013) Tools for wetland management: lessons learnt from a comparative assessment. *Environmental Science & Policy* (in press). doi:10.1016/j.envsci.2013.01.013

Deegan, L., Johnson, D.S., Warren, R.S., Peterson, B.J., Fleeger, J.W., Fagherazzi, S. and Wollheim, W.M. (2012) Coastal eutrophication as a driver of salt marsh loss. *Nature* 490, 388–392.

Department of Conservation (2007) The economic values of Whangamarino Wetland. Document No. DOCDM-141075, Department of Conservation New Zealand. Available at: http://www.doc.govt.nz/Documents/conservation/threats-and-impacts/benefits-of-conservation/economic-values-whangamarino-wetland.pdf (accessed February 2013).

Dudgeon, D., Arthington, A.H., Gessner, M.O., Kawabata, Z., Knowler, D.J., Lévêque, C., Naiman, R.J.,
   Prieur-Richard, A., Soto, D., Stiassny, M.L.J. and Sullivan, C.A. (2006) Freshwater biodiversity:
   importance, threats, status and conservation challenges. *Biological Reviews* 81, 163–182. doi:10.1017/
   S1464793105006950

Dugan, P., Sugunan, V.V., Welcomme, R.L., Béné, C., Brummett, R.E., Beveridge, M.C.M. *et al.* (2007) Inland
   fisheries and aquaculture. In: Molden, D. (ed.) *Water for Food, Water for Life: Comprehensive
   Assessment of Water Management in Agriculture.* Earthscan, London, in association with International
   Water Management Institute (IWMI), Colombo, pp. 459–483.

Dung, L.C., Hoanh, C.T., Le Page, C., Bousquet, F. and Gajeseni, N. (2009) Facilitating dialogue between
   aquaculture and agriculture; lessons from role playing games with farmers in the Mekong Delta,
   Vietnam. *Water Policy* 11(Supplement 11), 80–93. doi:10.2166/wp.2009.105

Elphick, C.S., Parsons, K.C., Fasola, M. and Mugica, L. (2010) Ecology and conservation of birds in rice fields:
   a global review. *Waterbirds* 33, Special Publication 1.

Falconer, I.R. (2001) Toxic cyanobacterial bloom problems in Australian waters: risks and impacts on human
   health. *Phycologia* 40 (3), 228–233. doi:10.2216/i0031-8884-40-3-228.1

Falkenmark, M., Finlayson, C.M., Gordon, L.J. *et al.* (2007) Agriculture, water, and ecosystems: avoiding the
   costs of going too far. In: Molden, D. (ed.) *Water for Food, Water for Life: Comprehensive Assessment of
   Water Management in Agriculture.* Earthscan, London, in association with International Water
   Management Institute (IWMI), Colombo, pp. 233–277.

FAO (2010) *The State of Food Insecurity in the World. Addressing Food Insecurity in Protracted Crises.* Food
   and Agriculture Organization of the United Nations, Rome.

FAO (2012) *The State of World Fisheries and Aquaculture 2012.* FAO Fisheries and Aquaculture Department,
   Food and Agriculture Organization of the United Nations, Rome. Available at: www.fao.org/docrep/016/
   i2727e/i2727e00.htm (accessed February 2013).

Finlayson, C.M. (2011) Managing aquatic ecosystems. In: Wilderer, P. (ed.) *Treatise on Water Science, Volume
   1.* Academic Press, Oxford, pp. 35–59.

Finlayson, C.M. and D'Cruz, R. (2005) Inland water systems. In: Hassan, R., Scholes, R. and Ash, N. (eds)
   (2005) *Ecosystems and Human Well-being: Current State and Trends, Volume 1. Findings of the
   Condition and Trends Working Group of the Millennium Ecosystem Assessment.* World Resources
   Institute and Island Press, Washington, DC, pp. 551–584.

Finlayson, C.M., Davidson, N.C., Spiers A.G. and Stevenson, N.J. (1999) Global wetland inventory – current
   status and future priorities. *Marine and Freshwater Research* 50, 717–727. doi:10.1071/MF99098

Foley, J.A. *et al.* (2005) Global consequences of land use. *Science* 309, 570–574. doi:10.1126/science.1111772

Folke, C., Kautsky, N., Berg, H., Jansson, A. and Troell, M. (1998) The ecological footprint concept for
   sustainable seafood production: a review. *Ecological Applications* 8, S63–S71. doi:10.1890/1051-
   0761(1998)8[S63:TEFCFS]2.0.CO;2

Frei, M. and Becker, K. (2005) Integrated rice–fish production and methane emission under greenhouse
   conditions. *Agriculture, Ecosystems and Environment* 107, 51–56.

Gichuki, F.N., Kodituwakku, D.C., Nguyen-Khoa, S. and Hoanh, C.T. (2009) Cross-scale trade-offs and
   synergies in aquaculture, water quality and environment: research issues and policy implications. *Water
   Policy* 11(Supplement 1), 1–12. doi:10.2166/wp.2009.100

Gregory, P.J. *et al.* (2002) Environmental consequences of alternative practices for intensifying crop
   production. *Agriculture, Ecosystems and Environment* 88, 279–290.

Gurung, T.B. (2002) Fisheries and aquaculture activities in Nepal. *Aquaculture Asia* 7, 39–44.

Hall, S.J., Delaporte, A., Phillips, M.J., Beveridge, M.C.M. and O'Keefe, M. (2011) *Blue Frontiers: Managing
   the Environmental Costs of Aquaculture.* WorldFish, Penang.

Hart, B.T., van Dok, W. and Djuangsih, N. (2002) Nutrient budget for Saguling Reservoir, West Java,
   Indonesia. *Water Research* 36, 2152–2160.

Hoanh, C.T., Tuong, T.P. and Hardy, B. (eds) (2006) *Environment and Livelihoods in Tropical Coastal Zones:
   Managing Agriculture–Fishery–Aquaculture Conflicts.* Comprehensive Assessment of Water
   Management in Agriculture Series 2. CAB International, Wallingford, UK in association with International
   Rice Research Institute (IRRI), Los Banos, Philippines and International Water Management Institute
   (IWMI), Colombo.

Hollis, G.E., Adams, W.M. and Kano, M.A. (eds) (1993) *The Hadejia-Nguru Wetlands: Environment, Economy
   and Sustainable Development of a Sahelian Floodplain Wetland.* IUCN, Gland, Switzerland.

Horwitz, P. and Finlayson, C.M. (2011) Wetlands as settings: ecosystem services and health impact assessment
   for wetland and water resource management. *BioScience* 61, 678–688.

Johnston, R., Cools, J., Liersch, S., Morardet, S., Murgue, C., Mahieu, M., Zsuffa, I. and Uyttendaele, G.P. (2013) WETwin: A structured approach to evaluating wetland management options in data-poor contexts. *Environmental Science & Policy* (in press). doi:10.1016/j.envsci.2012.12.006

Le Quesne, T. *et al.* (2010) *Flowing Forward: Freshwater Ecosystem Adaptation to Climate Change in Water Resources Management and Biodiversity Conservation.* Water Working Notes No. 28, The World Bank, Washington, DC.

Lehner, B. and Döll, P. (2004) Development and validation of a global database of lakes, reservoirs and wetlands. *Journal of Hydrology* 296, 1–22. doi:10.1016/j.jhydrol.2004.03.028

Little, D.C. and Bunting, S.W. (2005) Opportunities and constraints to urban aquaculture, with a focus on south and southeast Asia. In: Costa-Pierce, B.A., Edwards, P., Baker, D. and Desbonnet, A. (eds) *Urban Aquaculture.* CAB International, Wallingford, UK, pp. 25–44.

Lorenzen, K., Beveridge, M.C.M. and Mangel, M. (2012) Cultured fish: integrative biology and management of domestication and interactions with wild fish. *Biological Reviews* 87, 639–660.

Lukatelich, R.J. and McComb, A.J. (1986) Nutrient levels and the development of diatoms and blue–green algal blooms in shallow Australian estuary. *Journal of Plankton Research* 8, 597–618. doi:10.1093/plankt/8.4.597

Maclean, J.J., Dawe, D.C., Hardy, B. and Hettel, G.P. (eds) (2002) *Rice Almanac,* 3rd edn. International Rice Research Institute (IRRI), Los Baños, Philippines, West Africa Rice Development Association (WARDA), Bouaké, *Côte d'Ivoire* [now AfricaRice, presently at Cotonou, Benin], International Center for Tropical Agriculture (ICTA), Cali, Colombia and Food and Agriculture Organization, Rome.

Matsuno, Y., Ko, H.S., Tan, C.H., Barker, R. and Levine, G. (2002) *Accounting of Agricultural and Nonagricultural Impacts of Irrigation and Drainage Systems: A Study of Multifunctionality in Rice.* IWMI Working Paper 43, International Water Management Institute (IWMI), Colombo.

McCartney, M. and Smakhtin, V. (2010) *Water Storage in an Era of Climate Change: Addressing the Challenge of Increasing Rainfall Variability.* IWMI Blue Paper, International Water Management Institute, Colombo.

McCartney, M.P. and van Koppen, B. (2004) *Wetland Contributions to Livelihoods in United Republic of Tanzania.* FAO–Netherlands Partnership Programme: Sustainable Development and Management of Wetlands, Food and Agriculture Organization of the United Nations, Rome.

McCartney, M., Rebelo, L.-M., Senaratna Sellamuttu, S. and de Silva, S. (2010) *Wetlands, Agriculture and Poverty Alleviation: Appropriate Management to Safeguard and Enhance Wetland Productivity.* IWMI Research Report 137, International Water Management Institute, Colombo. doi:10.5337/2010.230

McCartney, M., Morardet, S., Rebelo, L-M. and Finlayson, C.M. (2011a) A study of wetland hydrology and ecosystem service provision: GaMampa Wetland, South Africa. *Hydrological Sciences Journal* 56, 1452–1466.

McCartney, M., Rebelo, L.-M., Mapedza, E., de Silva, S. and Finlayson, C.M. (2011b) The Lukanga Swamps: use, conflicts and management. *Journal of International Wildlife Law and Policy* 14, 293–310.

McInnes, R. (2010) Urban Development, Biodiversity and Wetland Management: Expert Workshop Report. UN Habitat and Ramsar Convention on Wetlands Expert Workshop, 16 to 17 November 2009, Kenya Wildlife Service Training Institute, Naivasha, Kenya. Bioscan UK), Oxford, UK. Available at: http://www.unhabitat.org/downloads/docs/ExpertWorkshopWetlands.pdf (accessed February 2013).

Millennium Ecosystem Assessment (2005) *Ecosystems and Human Well-being: Wetlands and Water – Synthesis. A Report of the Millennium Ecosystem Assessment.* World Resources Institute, Washington, DC. Available at: www.maweb.org/documents/document.358.aspx.pdf (accessed February 2013).

Mills, D.J., Westlund, L., de Graaf, G., Kura, Y., Willman, R. and Kelleher, K. (2010) Underreported and undervalued: small-scale fisheries in the developing world. In: Pomeroy, R.S. and Andrew, N. (eds) *Small-scale Fisheries: Management Frameworks and Approaches for the Developing World.* CAB International, Wallingford, UK, pp. 1–15.

Molden, D. (ed.) (2007) *Water for Food, Water for Life: Comprehensive Assessment of Water Management in Agriculture.* Earthscan, London, in association with International Water Management Institute (IWMI), Colombo.

Munro, J.L., Iskandar and Costa-Pierce B.A. (1990) Fisheries of the Saguling reservoir and a preliminary appraisal of management options. In: Costa-Pierce, B.A. and Soemarawoto, O. (eds) *Reservoir Fisheries and Aquaculture Development for Resettlement in Indonesia.* ICLARM Technical Report 23, Perusahaan Umum Listrik Negara, Jakarta, Indonesia, Institute of Ecology, Padjadjaran University, Bandung, Indonesia and International Center [Centre] for Living Aquatic Resources Management, Metro Manila, Philippines (ICLARM) [now WorldFish, Penang], pp. 285–328.

Nagabhatla, N., Beveridge, M.C.M., Nguyen-Khoa, S., Haque, A.B.M. and Van Brakel, M. (2012a) Multiple water use as an approach for increased basin productivity and improved adaptation: A case study from Bangladesh. *International Journal of River Basin Management*, 10, 121–136.

Nagabhatla, N., Dhyani, S., Finlayson, C.M., Senaratna Sellamuttu, S., Van Brakel, M., Wickramasuriya, R., Pattanaik, C. and Narendra Prasad, S. (2012b) A case study approach to demonstrate the use of assessment and monitoring tools for participatory environmental governance. *Ecologia* 2, 60–75.

Nguyen Thi, H.T., Nguyen Thi, T., Nguyen Hai, D., Do Van, T. and Nguyen Thi, D.P. (2010) *Situation Analysis Report on Highland Aquatic Resources Conservation and Sustainable Development in Northern and Central Vietnam*. HighARCS (Highland Aquatic Resources Conservation and Sustainable Development) Project Report, Research Institute for Aquaculture No. 1, Bac Ninh, Vietnam.

Pachauri, R.K. and Reisinger, A. (2007) *Climate Change 2007: Synthesis Report. Contribution of Working Groups I, II and III to the Fourth Assessment Report of the Intergovernmental Panel on Climate Change (IPCC)*. IPCC, Geneva, Switzerland.

Panigrahi, S., Acharya, B.C., Rama Chandra, P., Bijaya Ketan, N., Banarjee, K. and Sarkar, S.K. (2007) Anthropogenic impact on water quality of Chilika Lagoon RAMSAR site: a statistical approach. *Wetlands Ecology and Management* 15, 113–126.

Pauly, D. (1995) Anecdotes and the shifting baseline syndrome of fisheries. *Trends in Ecology and Evolution* 10, 430.

Peden, D., Ayalneh, W., El Wakeel, A., Fadlalla, B., Elzaki, R., Faki, H., Mati, B., Sonder, K. and Workalemahu, A. (2005) *Investment Options for Integrated Water-Livestock-Crop Production in Sub-Saharan Africa*. International Livestock Research Institute (ILRI), Addis Ababa.

Ramsar Convention Secretariat (2011) *The Ramsar Convention Manual: A Guide to the Convention on Wetlands (Ramsar, Iran, 1971)*, 5th edn. Ramsar Convention Secretariat, Gland, Switzerland.

Rebelo, L.-M., McCartney, M.P. and Finlayson, C.M. (2010) Wetlands of sub-Saharan Africa: distribution and contribution of agriculture to livelihoods. *Wetlands Ecology and Management* 18, 557–572. doi:10.1007/s11273-009-9142-x

Rebelo, L.-M., McCartney, M.P. and Finlayson, C.M. (2011) The application of geospatial analyses to support an integrated study into the ecological character and sustainable use of Lake Chilwa. *Journal of Great Lakes Research* 37, 83–92.

Senaratna Sellamuttu, S., de Silva, S., Nagabhatla, N., Finlayson, C.M., Pattanaik, C. and Prasad, N. (2012) The Ramsar Convention's wise use concept in theory and practice: an inter-disciplinary investigation of practice in Kolleru Lake, India. *Journal of International Wildlife Law and Policy* 15, 228–250.

Shiklomanov, I.A. and Rodda, J. (2003) *World Water Resources at the Beginning of the 21st Century*. United Nations Educational, Scientific and Cultural Organization (UNESCO), Paris.

Smith, L.E.D., Nguyen-Khoa, S. and Lorenzen, K. (2005) Livelihood functions of inland fisheries: policy implications in developing countries. *Water Policy* 7, 359–384.

Springate-Baginski, O., Allen, D. and Darwall, W.R.T. (eds) (2009) *An Integrated Wetland Assessment Toolkit: A Guide to Good Practice*. International Union for Conservation of Nature (IUCN), Gland, Switzerland and IUCN Species Programme, Cambridge, UK.

Szuster, B., Molle, F., Flaherty, M. and Srijantr, T. (2003) Socioeconomic and environmental implications of inland shrimp farming in the Chao Phraya Delta. In: Molle, F. and Srijantr, T. (eds) *Perspective on Social and Agricultural Change in the Chao Phraya Delta*. White Lotus Press, Bangkok, Thailand.

Tran Huu, T., Mai Van, X., Do Nam and Navrud, S. (2009) Valuing direct use values of wetlands: A case study of Tam Giang–Cau Hai lagoon wetland in Vietnam. *Ocean and Coastal Management* 52, 102–112. doi:10.1016/j.ocecoaman.2008.10.011

Troell, M., Joyce, A. and Chopin, T. (2009) Ecological engineering in aquaculture – potential for integrated multi-trophic aquaculture (IMTA) in marine offshore systems. *Aquaculture* 297, 1–9.

UNEP (2006) *Marine and Coastal Ecosystems and Human Well-being: A Synthesis Report Based on the Findings of the Millennium Ecosystem Assessment*. United Nations Environment Programme, Nairobi.

UNEP (2010) *Blue Harvest: Inland Fisheries as an Ecosystem Service*. WorldFish, Penang, Malaysia and United Nations Environment Programme, Nairobi.

Valiela, I., Bowen, J.L. and York, J.K. (2001) Mangrove forests: one of the world's threatened major tropical environments. *BioScience* 51, 807–815.

Welcomme, R.L., Cowx, I.G., Coates, D., Béné, C., Funge-Smith, S., Halls, A. and Lorenzen, K. (2010) Inland capture fisheries. *Philosophical Transactions of the Royal Society B* 365, 2881–2896. doi:10.1098/rstb.2010.0168

WHO (2006) *Guidelines for the Safe Use of Wastewater, Excreta and Greywater in Agriculture.* World Health Organization, Geneva, Switzerland.

Wood, A. and van Halsema, G.E. (2008) *Scoping Agriculture-Wetland Interactions. Towards a Sustainable Multiple Response Strategy.* FAO Water Reports 33, Food and Agriculture Organization of the United Nations, Rome.

World Bank, FAO and IFAD (2009) *Gender in Agriculture Sourcebook.* World Bank, Washington, DC with Food and Agriculture Organization of the United Nations, Rome and International Fund for Agricultural Development, Rome.

WRI, UNDP, UNEP and World Bank (2008) *World Resources 2008: Roots of Resilience: Growing the Wealth of the Poor – Ownership – Capacity – Connection.* World Resources Institute in collaboration with United Nations Development Programme, United Nations Environment Programme and World Bank. World Resources Institute, Washington, DC.

WWF (2009) Wetland restoration: Sand River, South Africa. WWF Africa (World Wildlife Fund for Nature) South Africa, Claremont, South Africa. Available at: wwf.panda.org/who_we_are/wwf_offices/south_africa/index.cfm?uProjectID=ZA0313 (accessed October 2010).

Xie, J., Hu, L., Tang, J., Wu, X., Li, N., Yuan, Y., Yang, H., Zhang, J., Luo, S. and Chen, X. (2011) Ecological mechanisms underlying the sustainability of the agricultural heritage rice–fish coculture system. *Proceedings of the National Academy of Sciences* 108, E1381–E1387. doi:10.1073/pnas.1111043108

8 Increasing Water Productivity in Agriculture

Katrien Descheemaeker,[1*] Stuart W. Bunting,[2] Prem Bindraban,[3]
Catherine Muthuri,[4] David Molden,[5] Malcolm Beveridge,[6]
Martin van Brakel,[7] Mario Herrero,[8] Floriane Clement,[9] Eline Boelee,[10]
Devra I. Jarvis[11]

[1]Plant Production Systems, Wageningen University, Wageningen, the Netherlands;
[2]Essex Sustainability Institute, University of Essex, Colchester, UK; [3]World Soil
Information (ISRIC) and Plant Research International, Wageningen, the Netherlands;
[4]World Agroforestry Centre (ICRAF), Nairobi, Kenya; [5]International Centre for
Integrated Mountain Development (ICIMOD), Kathmandu, Nepal; [6]WorldFish, Lusaka,
Zambia; [7]CGIAR Research Program on Water, Land and Ecosystems, 2075, Colombo,
Sri Lanka; [8]Commonwealth Scientific and Industrial Research Organisation (CSIRO),
St Lucia, Queensland, Australia; [9]International Water Management Institute (IWMI),
Kathmandu, Nepal; [10]Water Health, Hollandsche Rading, the Netherlands;
[11]Bioversity International, Rome, Italy

Abstract

Increasing water productivity is an important element in improved water management for
sustainable agriculture, food security and healthy ecosystem functioning. Water productivity is
defined as the amount of agricultural output per unit of water depleted, and can be assessed for
crops, trees, livestock and fish. This chapter reviews challenges in and opportunities for improving
water productivity in socially equitable and sustainable ways by thinking beyond technologies,
and fostering enabling institutions and policies. Both in irrigated and rainfed cropping systems,
water productivity can be improved by choosing well-adapted crop types, reducing unproductive
water losses and maintaining healthy, vigorously growing crops through optimized water, nutrient
and agronomic management. Livestock water productivity can be increased through improved
feed management and animal husbandry, reduced animal mortality, appropriate livestock
watering and sustainable grazing management. In agroforestry systems, the key to success is
choosing the right combination of trees and crops to exploit spatial and temporal complementarities
in resource use. In aquaculture systems, most water is depleted indirectly for feed production, via
seepage and evaporation from water bodies, and through polluted water discharge, and efforts to
improve water productivity should be directed at minimizing those losses. Identifying the most
promising options is complex and has to take into account environmental, financial, social and
health-related considerations. In general, improving agricultural water productivity, thus freeing
up water for ecosystem functions, can be achieved by creating synergies across scales and
between various agricultural sectors and the environment, and by enabling multiple uses of water
and equitable access to water resources for different groups in society.

* E-mail: katrien.descheemaeker@wur.nl

Background

As water resources around the world are threatened by scarcity, degradation and overuse, and food demands are projected to increase, it is important to improve our ability to produce food with less water. There are only a few basic methods of using the earth's water resources to meet the growing food demands: continuing to expand rainfed and irrigated lands; increasing production per unit of water; trade in food commodities; and changes in consumption practices. Land expansion is no longer a viable solution (Godfray *et al.*, 2010). Therefore, improving agricultural productivity on existing lands using the same amount of water will be essential. Increasing water productivity means using less water to complete a particular task, or using the same amount of water, but producing more. Increased water productivity has been associated with improved food security and livelihoods (Cook *et al.*, 2009b; Cai *et al.*, 2011). Additionally, it leads to savings in fresh water, making it available for other uses, such as healthy ecosystem functioning. Increased water productivity is therefore an important element in improved management of water and ecosystems for sustainable agriculture and food security.

Water productivity is the amount of beneficial output per unit of water depleted. In its broadest sense, it reflects the objectives of producing more food, and the associated income, livelihood and ecological benefits, at a lower social and environmental cost per unit of water used (Molden *et al.*, 2007). Usually, water productivity is defined as a mass (kg), monetary ($) or energy (calorific) value of produce per unit of water evapotranspired (Kijne *et al.*, 2003; Molden *et al.*, 2010), and, as such, it is a measure of the ability of agricultural systems to convert water into food. Water use efficiency and water productivity are often used in the same context of increasing agricultural outputs while using or degrading fewer resources. Although definitions vary, water use efficiency usually takes into account the water input, whereas water productivity uses the water consumption in its calculation. In this chapter, both terms are used interchangeably, reflecting the most common use in a specific field.

Improving agricultural water productivity is about increasing the production of rainfed or irrigated crops, but also about maximizing the products and services from livestock, trees and fish per unit of water use. Crop water productivity has been the subject of many years of research, and its assessment and means for improvement are well documented (Kijne *et al.*, 2003; Bouman, 2007; Molden, 2007; Rockström and Barron, 2007). However, for other agricultural outputs and systems, such as livestock, agroforestry, fisheries and aquaculture, research on improving water productivity is still in its infancy. In recent years though, a growing body of evidence is creating a clearer picture on the potential solutions and ways forward (Cai *et al.*, 2011). Besides going beyond crops, this chapter also emphasizes the need for careful targeting of technologies and enabling policies and institutions for successful adoption in farmer communities. Other cross-sectoral approaches for improved water productivity, such as multiple use of water, reducing postharvest losses and basin studies will be discussed briefly.

Increasing Crop Water Productivity

Opportunities for improving crop water productivity mainly lie in choosing adapted, water-efficient crops, reducing unproductive water losses and ensuring ideal agronomic conditions for crop production (see, for example, Kijne *et al.*, 2003; Bouman, 2007; Rockström and Barron, 2007). In general, agronomic measures directed at healthy, vigorously growing crops favour transpirational and productive water losses over unproductive losses. An important principle for crop water productivity is that taking away water stress will only improve water productivity if other stresses (nutrient deficiencies, weeds and diseases) are also alleviated or removed (Bouman, 2007), i.e. water management should go hand in hand with nutrient manage-ment, soil management and pest management (Bindraban *et al.*, 1999; Rockström and Barron, 2007). Since the Comprehensive Assessment of Water Management in Agriculture, of which the main ouput was the report *Water for Food, Water for Life* (Molden, 2007), research on the performance of various interventions for crop water productivity improvement has included, among others, supplemental irrigation, precision irrigation and drainage, soil fertility management, reduced tillage operations, soil moisture conservation, and the use of drought- and disease-resistant crop varieties (Fischer *et al.*, 2009; Geerts and Raes, 2009; Gowda *et al.*, 2009; Oweis and Hachum, 2009a,b; Stuyt *et al.*, 2009; de Vries *et al.*, 2010; Arora *et al.*, 2011; Balwinder *et al.*, 2011; Mzezewa *et al.*, 2011).

There is great variation in water productivity across cropping systems, under both irrigated and rainfed conditions. It has been estimated that three quarters of the additional food we need for our growing population could be met by increasing the productivity of low-yield farming systems, probably to 80% of the productivity that high-yield farming systems obtain from comparable land (Molden, 2007). Especially where yield gaps are large, there is large scope for improvement (de Fraiture and Wichelns, 2010; Cai *et al.*, 2011). In that respect, the highest potential water productivity gains can be achieved in low-yielding rainfed areas in pockets of poverty across much of sub-Saharan Africa and South Asia (Rockström *et al.* 2010). As many of the world's poorest people live in currently low-yielding rainfed areas, improving the productivity of water and land in these areas would result in multiple benefits. Thus, by getting more value out of currently underutilized rainwater, agricultural land expansion would be limited, and the livelihoods of these poor men and women would be improved, without threatening other ecosystem services (WRI *et al.*, 2008).

A recent global analysis on closing yield gaps indicated that appropriate nutrient and water management are essential and have to go hand in hand (Mueller *et al.*, 2012). Comparing bright spots (examples of high water productivity) with hot spots (examples of low water productivity) across ten different basins showed that yield increases through tailored interventions are possible at many locations and would lead to major gains in water productivity (Cai *et al.*, 2011). Gaps in crop water productivity are often linked to

access to water, but also to access to other inputs such as seeds and fertilizers, which illustrates the importance of markets and infrastructure (Ahmad and Giordano, 2010). However, in highly productive areas, caution on the scope for gains in crop water productivity is warranted (Molden *et al.*, 2010). There is a crop-dependent biophysical limit to the biomass production per unit of transpiration (Seckler *et al.*, 2003; Steduto *et al.*, 2007; Gowda *et al.*, 2009), and whereas plant breeders have managed to increase the harvest index of crops (the ratio of marketable produce to total biomass), gains in this index appear to have peaked (Molden *et al.*, 2010). The canopy development that is associated with increasing yields limits the scope for reducing water losses, because doubling the yield also requires almost twice the amount of transpiration.

Increasing Water Productivity in Agroforestry Systems

The area under agroforestry worldwide was estimated at 1023 million ha in 2009, but it has been suggested that substantial additional areas of unproductive crop, grass and forest lands, as well as degraded lands, could be brought under agroforestry (Nair *et al.*, 2009). The concept of agroforestry is based on the premise that structurally and functionally more complex land use systems capture resources more efficiently than monocultures (Schroth and Sinclair, 2003). Agroforestry enhances resource utilization by improving temporal and/or spatial complementarity in resource capture (Ong *et al.*, 2007). Trees enhance below-ground diversity and this supports local ecosystem stability and resilience (Barrios *et al.*, 2012); trees also provide connectivity with forests and other features at the landscape and watershed levels (Harvey *et al.*, 2006). Agroforestry provides numerous benefits, ranging from diversification of production to improved exploitation of natural resources and provision of environmental functions, such as soil conservation (protection against erosion), improvement or maintenance of soil fertility, water conservation and more productive use of water (Cooper *et al.*, 1996).

Trees outside forests, or trees on farms, are an important component of man-made landscapes. With 10% tree cover on nearly half of the world's agricultural land, agroforestry is a common reality (Zomer *et al.*, 2009). Trees are important landscape elements that help regulate water flows. Even a small change in tree cover can have a large impact on reducing runoff and enhancing infiltration and transpiration (Carroll *et al.*, 2004; Hansson, 2006), through the use of the trees to provide fuelwood, fodder, fruit and timber (Ong and Swallow, 2003). 'Hydraulic lift' is an interesting phenomenon in agroforestry systems, whereby the tree root system lifts water from moist deep soil layers to the upper soil layers, where it is accessible to crops (Roupsard, 1997; Ong and Leakey, 1999; Bayala *et al.*, 2008). Agroforestry belts have also been proposed as riparian buffers to combat non-point source water pollution from agricultural fields and help to clean runoff water by reducing runoff velocity, thereby promoting infiltration, sediment deposition and nutrient retention (Jose, 2009). The management of riparian vegetation can improve the quality of water in the river and hence, via its outflow, help to protect valuable coastal ecosystems, such as the Great Barrier Reef (Pert *et al.*, 2010). In degraded areas of the Abay Basin in Ethiopia, integrating multi-purpose trees into farms helped to fight land degradation while increasing the productive use of water (Merrey and Gebreselassie, 2011).

A key challenge for agroforestry is to identify which combination of tree and crop species optimizes the capture and use of scarce environmental resources such as light, water and nutrients, at the same time as fulfilling farmers' needs for timber, fuel, mulch, fodder and staple food (Sanchez, 1995; Muthuri *et al.*, 2009). The complementary aspects of trees in relation to crops can be enhanced by selecting and managing trees to minimize competition (Schroth, 1999) by means of root and shoot pruning (Siriri *et al.*, 2010), increasing tree spacing within the crops (Singh *et al.* 1989), and matching the trees and crops to appropriate niches within the farm (van Noordwijk and Ong, 1996).

Increasing Livestock Water Productivity

Livestock products provide one third of the human protein intake, but also consume almost one third of the water used in agriculture globally (Herrero *et al.*, 2009). Most of the world's animal production comes from rainfed mixed crop–livestock systems in developing countries and from intensive industrialized production in developed countries (Herrero *et al.*, 2010). Livestock production systems are rapidly changing in response to various drivers, which calls for the constant adaptation of policy, investment and technology options (Chapter 2). With increasing demands for animal products, along with increasing global water scarcity and competition for water, improving livestock water productivity (LWP) has become essential (Descheemaeker *et al.*, 2010a).

LWP was first defined by Peden *et al.* (2007) as the ratio of livestock products and services to the water depleted and degraded in producing these; it can also include water depleted in slaughterhouses and milk-processing facilities. Since the launch of the LWP concept, several studies have investigated the livestock–water nexus and dealt with LWP at various scales (Amede *et al.*, 2009a,b; Cook *et al.*, 2009a; Gebreselassie *et al.*, 2009; Haileslassie *et al.*, 2009a,b; van Breugel *et al.*, 2010; Descheemaeker *et al.*, 2011; Mekonnen *et al.*, 2011). While offering good insights into how LWP can be increased, these studies have also advanced the methodologies for LWP assessment. A remaining question is how to account for the value of the water consumed (Peden *et al.*, 2009b). For example, livestock grazed on arid and semi-arid pastures utilize water that cannot be used for crops and would be depleted through evapotranspiration before it could enter groundwater and surface water bodies (Bindraban *et al.*, 2010). Such water would be valued less than water in an irrigation scheme that can be used for growing high-value vegetable crops. A consideration of the value of water could lead to demand-side management that would foster a rebalancing of water use among agricultural sectors. Especially for livestock production in areas of low potential and in smallholder systems, such considerations would show that livestock are very efficient in making productive use of water that is of low value for other sectors.

Global environmental evidence suggests that the livestock sector has a strong negative impact on water depletion and pollution (Steinfeld et al., 2006). However, caution is needed with respect to such pronouncements, because big differences exist between various livestock systems and agroecologies. For example, in industrial livestock systems, soil and water contamination from manure and wastewater mismanagement and the use of chemicals is a common problem, whereas in smallholder low-input systems this is not (yet) the case. In these smallholder systems, livestock often provide multiple services, including farm power for cultivation and transport, and manure for soil fertility management (Tarawali et al., 2011). Valuing manure as a beneficial output of livestock systems would result in a much higher figure for LWP than when only meat and milk are taken into account. This illustrates the importance of the context in which livestock productivity assessments are made (Cai et al., 2011).

Calculations of LWP have shown that servicing and drinking, though at first sight the most obvious water uses of livestock, in reality constitute only a minor part of the total water consumption in livestock-based agroecosystems (Peden et al., 2007, 2009a). The major water depletion in relation to livestock production is the evapotranspiration of water for feed production (Peden et al., 2007; Gebreselassie et al., 2009). The large global variations in feed water productivity (see Table 8.1) are not only a sign of divergent methodologies, but also illustrate that LWP depends on the type, the growing conditions and the management of forage production. Hence, the large variation in LWP in the Nile Basin (Box 8.1) is not surprising, and illustrates that there is ample scope for improvement.

Innovative interventions for improved LWP can be grouped in three categories (Peden et al., 2009b; Descheemaeker et al., 2010a; Herrero et al., 2010):

- Feed-related strategies for improving LWP comprise: the careful selection of feed types, including crop residues and other waste products; improving the nutritional quality of the feed; optimizing the use of multi-purpose food–feed–timber crops; increasing feed water productivity by appropriate crop and cultivar selection and improved agronomic management; and implementing more sustainable grazing management practices.
- Water management strategies for higher LWP consist of water conservation and water harvesting, strategic placement and monitoring of watering points, and the integration of livestock production into irrigation schemes.

Table 8.1. Global ranges of feed water productivity for different feed types, derived from the literature.[a]

Feed type	Feed water productivity (kg/m^3)
Cereal grains	0.35–1.10
Cereal forages	0.33–2.16
Food–feed crops (total biomass)	1.20–4.02
Irrigated lucerne	0.80–2.30
Pastures	0.34–2.25
(Semi)-arid rangelands	0.15–0.60

[a]Ferraris and Sinclair, 1980; Sala et al., 1988; Bonachela et al., 1995; Saeed and El-Nadi, 1997, 1998; Renault and Wallender, 2000; Chapagain and Hoekstra, 2003; Oweis et al., 2004; Singh et al., 2004; Smeal et al., 2005; Nielsen et al., 2006; Gebreselassie et al., 2009; Haileslassie et al., 2009a,b; van Breugel et al., 2010.

Box 8.1. Livestock water productivity (LWP) in the Nile Basin

A basin-wide assessment of livestock water use and productivity showed that the total water need for feed production in the Nile Basin was roughly 94 billion m^3, which amounts to approximately 5% of the total annual rainfall (68 billion m^3, or 3.6% of total annual rainfall when excluding water for crop residues) (van Breugel *et al.*, 2010). In most areas of the basin, LWP is less than 0.1 US$/$m^3$, with only a few areas showing an LWP of 0.5 US$/$m^3$ and higher (Fig. 8.1). Livestock water productivity is on average low, but large differences exist across the basin, both within and between livestock production systems. These differences suggest that there is scope for improvement of LWP (see main text for an overview of options), which could lead to significant reduction of water use at the basin level while maintaining current levels of production. In line with the large-scale (basin-wide) analysis, community and household level analyses indicated that in the Ethiopian highlands, LWP ranges from 0.09 to 0.69 US$/$m^3$ (Haileslassie *et al.*, 2009b; Descheemaeker *et al.*, 2010b), whereas in animal feeding trials LWP ranged from 0.27 to 0.64 US$/$m^3$ (Gebreselassie *et al.*, 2009).

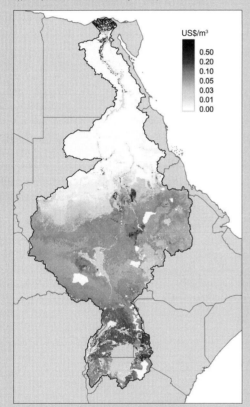

US$/$m^3$

0.50
0.20
0.10
0.05
0.03
0.01
0.00

Fig. 8.1. Livestock water productivity of the Nile Basin (outlined area) expressed as the ratio of the summed value of meat and milk and the water depleted to produce the required livestock feed. Water for residues was not included in the calculation of depleted water (Map by P. van Breugel, based on van Breugel *et al.*, 2010).

When considering just milk production, smallholder production systems in the Ethiopian highlands are characterized by very low water productivity, ranging between 0.03 and 0.08 l milk/m^3 (Descheemaeker *et al.*, 2010b; van Breugel *et al.*, 2010). In other words, the virtual water content of milk in these systems ranges from 12.5 to 33 m^3 water/l milk, which is very high considering the global average of 0.77 m^3 water/l milk (Chapagain and Hoekstra, 2003). However, the difference from the highly specialized and efficient industrial systems is that in smallholder systems, milk production is often viewed as a by-product of livestock keeping. Livestock are kept for multiple purposes and services (Thornton and Herrero, 2001; Moll *et al.*, 2007; Cecchi *et al.*, 2010), of which manure and draft power are usually more important than milk and meat production. The LWP concept and framework developed by the International Water Management Institute (IWMI) and International Livestock Research Institute (ILRI) (Peden *et al.*, 2007; Descheemaeker *et al.*, 2010a) allow the taking into account of these multiple livestock products and services in water productivity assessments.

- Animal management strategies include improving breeds, disease prevention and control, and appropriate animal husbandry, supported by raising awareness among livestock keepers that the same benefit can be obtained from smaller and fewer, but more productive, herds.

Designing LWP interventions that benefit the poor requires an understanding of the differentiated access to livestock-related capitals and livelihood strategies of men and women and of different socio-economic groups within local communities (Clement et al., 2011). Livestock often provide an important source of income for women, particularly in mixed crop–livestock systems. Furthermore, in order to facilitate their adoption, technological interventions need to be supported by appropriate policies and institutions (Amede et al., 2009b). For example, establishing institutions such as water users' associations, together with policies such as cost recovery for water use, can contribute to improving the efficiency of feed crop irrigation.

The important role of informal arrangements in LWP should not be underestimated as these can provide socially acceptable ways for different groups in society to access water (Adams et al., 1997). In communal grazing lands, for example, it is not only vegetation but also water resources that bind herders together, and arrangements are needed to ensure equitable access and sustainable use. Opportunities for the sustainable management of livestock grazing systems in a way that maintains ecosystem services include institutions that enable the management of climate variability – such as early warning and response systems, improved markets, livestock loss insurance schemes and fodder reserves (World Bank, 2009). Other approaches deal with changing the incentive system for keeping large herds, such as payment for environmental services and increasing the level of cost recovery in the use of natural resources, and veterinary services (World Bank, 2009). Such incentive systems require great attention to issues of equity and legitimacy, as they might increase existing or create new social inequities.

Increasing Water Productivity in Aquaculture

Benefits from aquaculture include the production of food, improved livelihoods, nutrition and health (Dugan et al., 2007). The abstraction and discharge of water for aquaculture may, however, affect ecological processes and compromise ecosystem services that support other livelihoods. Appropriation of water for aquaculture may lead to competition with other resource users, including other aquaculture operators. Water requirements for aquaculture are both qualitative and quantitative in nature, but the definition of the water quantities 'used' presents difficulties (Nguyen-Khoa et al., 2008). Consumptive use of water for the accumulation of aquatic resources biomass is negligible in aquaculture. The water is mainly consumed indirectly in the production of aquaculture feed or via percolation, seepage, and evaporation from ponds and stocked reservoirs. Water productivity can thus be defined as the mass or value of the aquaculture produce divided by the amount of water required for feed plus the amount of evaporation and seepage from the pond or reservoir.

Water productivity assessment in cage or pen aquaculture presents yet another challenge. Cages allow natural water exchange and, like capture fisheries, do not induce significant water losses to the system. The disadvantage is that cage aquaculture discharges large quantities of nutrients and metabolites directly to its aquatic environment. Hence, the relative environmental impact per ton of product of cage and pen aquaculture in inland waters is much higher than that of any other aquatic production system (Hall et al., 2011). Water use efficiency varies markedly between different aquaculture production systems (Table 8.2), although fish and crustaceans are more efficient than terrestrial animals in terms of feed-associated water use. However, on-farm use of non-feed associated water in aquaculture can be very high, attaining up to 45m^3 per kg produced in ponds.

Pressures to enhance water productivity in aquaculture (Box 8.2) derive from global changes and domain-specific challenges such as production efficiency, risk management,

Table 8.2. Water use efficiency (in m^3 water/kg fresh weight) in aquaculture systems (adapted from Bunting, 2013).

Aquaculture system	Water use efficiency	Water management characteristics
Traditional extensive fish pond culture	45[a]	Rainwater and drainage water are routinely channelled into fish ponds to compensate for seepage and evaporation losses; excessive water exchange is detrimental as it is desirable to retain nutrients within the pond
Flow-through ponds	30.1[a]	Water exchange of 20% of the pond volume/day removes waste and replenishes oxygen levels; annual production of 30 t/ha is attainable, but seepage and evaporation contribute to water loss in the system
Semi-intensive fish ponds	11.5[a]	Fish ponds fed with formulated pellet feed can yield 6 t/ha, while producing two crops annually, and with complete drainage to facilitate harvest; one fifth of water consumption is associated with feed inputs
Wastewater-fed aquaculture	11.4[b]	Wastewater is routinely fed into fish ponds in the East Kolkata Wetlands (West Bengal, India) to make up the water to a desirable level; estimates suggest 550,000 m^3/day of wastewater is used to produce 18,000 t/year of fish in 3900 ha of ponds
Intensively managed ponds	2.7[a]	Lined ponds provide an annual production of 100 t/ha, while intensive mixing results in evaporation of 2000 mm/year
Super-intensive recirculation systems	0.5–1.4[a]	Process water is recirculated with pumps and treated with mechanical filters, biofilters and disinfection technology; stocked animals are entirely dependent on high-protein formulated feed inputs

[a]Based on Verdegem *et al.*, 2006; [b]from Bunting, 2007.

conflict avoidance, legislation and controls, consumer demand and public perception (Verdegem *et al.*, 2006; Chapter 2). The water productivity of aquaculture can be increased through improving system design, good management, good water quality, good brood stock, or using a combination of non-competing species that fill different niches in the aquatic ecosystem. Practices and policies that include construction, systems design and operation, optimization of production efficiency, water management practices, horizontally integrated aquaculture systems (Box 7.1, Chapter 7), water rates and pollution taxes, and policy and planning have been identified as potential areas where water use efficiency in aquaculture could be improved. The integration of aquaculture with other agricultural and water uses has potential for enhancing the productivity of appropriated freshwater resources in a wider systems context. Reservoir storage water, for example, is usually committed to uses other than fish production, but fish can be stocked in these for complementary production, while making non-depletive use of water (Chapter 7).

Aquaculture producers have an interest in reducing the financial as well as the environmental costs of managing (regulating, moving and conditioning) water resources. Consequently, aquaculture farmers are generally active in trying to make more efficient use of appropriate water resources, and work hard to comply with discharge standards, whether statutory or imposed by the community. Moreover, on-farm water movement and wastewater discharge may increase the likelihood of stock escaping, resulting in

---

Box 8.2. Pressures inducing enhanced water use efficiency in aquaculture.

Pressures to enhance water productivity in aquaculture come from internal drives for production efficiency and management optimization, efforts to reduce risks and avoid conflict, obligations to comply with legislation and standards, and endeavours to assure consumers and bolster public perception (see Bunting, 2013).

Producers wish to limit the costs of appropriating, handling, conditioning and treating water, reduce production-enhancing resources lost from culture systems and avoid the liabilities and negative perceptions associated with discharging wastewater. Operators are conscious of the risks from disease, pests, predators and pollution that may be entrained in water appropriated for aquaculture. Water transfers and discharges increase the risk of stock escaping and causing negative environmental impacts and financial losses. Rising costs for fuel and feed, and new and emerging hazards, are prompting producers to become less reliant on externalizing technology and to adopt more extensive and diversified production strategies. Abstraction and wastewater discharges can cause negative environmental impacts and disrupt ecosystem services that sustain the livelihoods of others, thus giving rise to grievances and, potentially, to conflict. Failure to comply with legislation and standards concerning wastewater discharge standards may result in financial penalties for producers, while the imposition of charges for water use and effluent releases may prove prohibitive. Unfavourable commentary and media coverage on water use for aquaculture can result in local opposition, and negative perceptions among consumers may adversely affect demand for aquaculture products.

---

revenue loss and negative environmental impacts. Farmers also have an interest in reducing water intake, as this will lessen competition between various aquaculture producers, and help to avoid conflict with other water (and land) users.

In order to have marketable products, aquaculture producers must also manage animal health risks associated with their own water intake, which may be polluted, and also with the ingress of entrained aquatic organisms that may harbour pests and pathogens. Control measures adopted by farmers include screening inflows to prevent predators and other aquatic animals from entering, and restricting the abstraction of water as far as possible, depending instead on reducing stocking densities and promoting ecological processes to condition culture water for continued use.

Transition by producers to more intensive water management through mechanical pumping and aeration can further reduce dependence on the appropriation of natural water resources, but may exacerbate environmental problems associated with fuel extraction or electricity generation and greenhouse gas emissions. The comprehensive life cycle assessment (LCA) of aquaculture systems permits the identification of the least environmentally damaging production strategies. Further research and development are needed

to develop practical approaches to evaluating, in concert, the environmental and social (including gender) impacts, livelihoods outcomes, financial viability, and economic and ethical implications of aquaculture developments. In the short term, these assessments could make life harder for poor aquaculture farmers, with new costs for licences, rents and taxes. In the longer term, they may benefit as stricter controls can protect the ecological status of receiving water bodies and thereby secure water resources for other and future users. This would also maintain and enhance the stocks and flows of ecosystem services. Product and livelihood diversification should be looked at as well so as to reduce dependence on aquaculture and generate more regular cash flows and higher revenues.

Water Productivity and Fisheries

Capture fisheries in lakes, rivers and wetlands present a special case for water productivity assessment, and the use of the concept is relatively new in this area. The values and livelihood benefits are high, but often ignored or underestimated (Béné et al., 2010). Lemoalle (2008) and Brummett et al. (2010) argue that the concept of water productivity cannot be extended from managed systems, including aquaculture, to natural systems, including fisheries, for the purposes of attributing relative value and prioritizing water allocation. This is because: (i) fisheries do not induce any water losses to the system other than water incorporated in the harvested product; (ii) there is a difficulty in fully parameterizing fisheries ecology models; and (iii) the water productivity concept does not sufficiently capture inherent trade-offs between different uses of water (Nguyen-Khoa *et al.*, 2008). The term 'marginal water productivity', which represents the economic, social and other values lost when fisheries are affected by other developments in a watershed, is proposed as a more appropriate measure of water productivity in this system. However, the differences in benefits accrued from fisheries and agriculture, and the difficulties in determining ecosystem flows, make inter-sectoral comparisons difficult. If the objective of such a comparison is to support water allocation decisions, it needs to be acknowledged that both the water productivity and the marginal water productivity of fisheries compare poorly with the water productivity of cultivated crops (Brummett *et al.*, 2010).

An additional focus needs to be put on fisheries management, which is often difficult (Andrew *et al.*, 2007). Badly managed fisheries can compromise the physical integrity of aquatic environments through destructive gear use – a problem associated with the use of dredges and bottom trawls in marine environments – and through overfishing, which, ultimately, can reduce the economic value of provisioning (i.e. fish catches) and other ecosystem services.

The Role of Technologies, Policies and Institutions

Agriculture is done by people in communities and landscapes that host a variety of agroecological and socio-economic conditions. With such complexity, it is not surprising that prescribed technologies, for instance to increase water productivity, do not always work, or are abandoned by farmers who do not benefit from them (see also Chapter 9). Commonly, this is caused by inappropriate

targeting of technologies (e.g. Merrey and Gebreselassie, 2011). This can be improved by considering development domains (e.g. Kruseman *et al.*, 2006), which combine agricultural biophysical potential with economic and demographic factors. In addition, technological innovations are not gender neutral, and the neglect of gender and caste, class, or ethnic or religious differentiation within communities can reinforce existing inequities in access to and control over water. This can result in high environmental, health and social costs, such as chronic under-nutrition, decreased yields or loss of livelihood opportunities (Zwarteveen, 1995). A bad example of such neglect comes in the case where women are the main users of water, e.g. for vegetable production, but only men are trained for the operation and maintenance of technologies – which fall under the perceived 'male domain' (Berejena *et al.*, 1999).

In addition, many new technologies aimed at making water more accessible or cheaper, lead to higher water consumption and negative environmental consequences (Molden, 2007). There are many examples of upstream water users improving local productivity but utilizing so much water that little is left for downstream users (Molle *et al.*, 2010). In many areas, the large growth in the use of water pumps has led to water overuse and the decline of environmental flows and groundwater tables (Shah, 2009). This problem is worse where the use of agrochemicals has resulted in poor water quality (Falkenmark and Molden, 2008; UNEP, 2010). These challenges related to improved water access illustrate the importance of the co-implementation of water resource development on the one hand and of supporting regulations and policies on the other hand, in order to preserve both the quantity and quality of water resources.

The development of water infrastructure has been identified as a key strategy towards poverty reduction (World Bank, 2008; Kandiero, 2009). Such water infrastructure developments would include water supply and sanitation systems, and dam construction, as well as investments in irrigation (World Bank, 2008). Stakeholders may need guidance on how to develop appropriate infrastructure with a view to maximizing ecosystem services and reaching an equitable share of benefits between men and women, and among different social groups. The choice that stakeholders face is not only one of whether to build or not, but also how to build and how to integrate the multiple needs, interests and perceptions of local communities. Some of the older existing infrastructure needs rehabilitation and this could be done in such a way that it not only helps to reduce poverty by providing wider and more equitable access to water, but also reduces water losses in current distribution networks, improves the overall efficiency of water use networks, and caters for the wider agroecosystem and its various functions and services. Infrastructure projects, combined with new technological advancements, can create more efficient irrigation systems that lose less water to evapotranspiration. New technology for improving water efficiency, such as drip irrigation, biotechnology advances, improved pump technology and better water practices, is already in place in many areas of high productivity, and could be implemented in areas of lower productivity too.

The economic aspects of water management interventions need to be considered as well. If the initial investment cost, the return on investment and the effect on production risk and labour inputs are unfavourable, farmers are unlikely to adopt the intervention. Many studies have investigated the economic aspects of different irrigation and drainage options (Al-Jamal et al., 2001; Mintesinot et al., 2004; Nistor and Lowenberg-DeBoer, 2007; Capra et al., 2008; Hagos et al., 2009; Amarasinghe et al., 2012) and rainwater management options (Goel and Kumar, 2005; Merrey and Gebreselassie, 2011). However, generalized conclusions on the economic performance of different options are impeded by its case- and situation-specific nature.

Some solutions for improving water productivity lie outside the water sector, such as in markets, prices and subsidies, but these are hard to influence, as trade is conducted for many economic and strategic reasons, with water often last on the long list of reasons for trade (Wichelns, 2010). There are also serious questions about whether trade or food aid is a viable pathway to food security for places like sub-Saharan Africa. Some countries would

rather invest their resources in utilizing their water resources better, in order to produce their own food, and aim for greater food self-sufficiency and a reduction in trade. Countries can also focus on producing crops that do not require a lot of water, such as the small grains produced in sub-Saharan Africa. The implication is that we will probably have to rely on better agricultural practices, as suggested in this chapter. Nevertheless, trade will grow in importance, both in terms of rural–urban connections and internationally, as its impact on ecosystem services at production points and at consumption locations also grows (Chapter 2). Though the negative impacts of depleted water are likely to be disconnected from consumers, pricing changes, brought about by depleted water, might eventually influence consumption patterns.

Finally, the failure of technical interventions is usually related to the neglect of the necessary underpinning policies and institutions (Merrey and Gebreselassie, 2011). For example, the root cause of the poor performance of irrigation systems is often poor governance and management, inappropriate policies and availability of inputs, and subsidies of fertilizer or output prices (Mukherji et al., 2009). Simultaneously, technology development and related investments in other sectors may have far-reaching impacts on the water sector (Box 8.3; see also Chapter 2).

Bridging Scales and Water Management Concepts

A shift in thinking about water resource development and management is imperative, including bridging the strict division between rainfed and irrigated agriculture (Rockström et al., 2010). It would help to think of rain as the ultimate source of water for all agro-ecosystems, and consider agricultural water management options across a wide spectrum that includes large-scale gravity irrigation, small-scale irrigation systems, provision of supplemental irrigation, use of groundwater, demand management, water harvesting techniques, soil moisture storage, and conservation and drainage. Water storage options along the continuum from soil and groundwater to

Box 8.3. The link of the water sector with renewable energy developments.

Renewable energy developments show promise for reducing both the carbon and water footprints of energy production. However, the push for renewable energy can have significant impacts on water availability through, for example, the disruption of water flows by hydropower dams and higher water consumption in the production of biofuels (UNEP, 2007). In closed basins, such as in the western USA or in much of Europe, the hydropower potential has been exhausted (WWAP, 2009), but in the developing world, more large dams are likely to be constructed. Dams change the hydrological cycle and often have negative environmental effects, including the disruption of migratory fish production (e.g. Dugan *et al.*, 2010). Conversely, renewable technologies, such as biogas and solar power, may reduce the use of water for power generation: coal uses about 2 m^3 water/MW h of electricity produced, nuclear power 2.5 m^3 and petroleum 4 m^3 (WWAP, 2009). Extracting oil also uses lots of water – up to 45 m^3/MW h from tar sands, one of the largest 'new' sources of oil (WWAP, 2009). In contrast, the increased applications for biofuel have led to high demand, with significant impacts on and trade-offs for water use, food security and agroecosystems (e.g. Berndes, 2002; de Fraiture *et al.*, 2008; FAO, 2008, 2009; Hellegers *et al.*, 2008; Bindraban *et al.*, 2009).

natural wetlands and dams can make water more accessible at different spatial and temporal scales (McCartney and Smakhtin, 2010). These scales range from field and farm to the level of large dams serving various communities, and from year-round accessibility to bridging shorter or longer dry spells (Johnston and McCartney, 2010; Merrey and Gebreselassie *et al.*, 2011).

When moving between scales, the concept of water wastage can change. For example, when considering irrigation efficiencies, which usually turn out to be disappointingly low (e.g. Calzadilla *et al.*, 2008, revealed a range in irrigation efficiency from 40 to 70%), one may conclude that a lot of water is wasted. However, this conclusion overlooks the fact that farmers living in or near irrigation systems in water-scarce environments make ample reuse of drainage water. Much of the 'wasted' water can be important for home gardens (Molle and Renwick, 2005), livestock (Peden *et al.*, 2005), fish (Nguyen-Khoa *et al.*, 2005), domestic uses leading to improved health (Boelee *et al.*, 2007), or recharging aquifers. This is in line with the finding that multiple use of water by both men and women can greatly increase the total value of beneficial outputs per water unit used and hence increase productivity (Meinzen-Dick, 1997; Bouma *et al.*, 2011). Multiple use of water can be considered at landscape and basin level, where water is used for various purposes, including non-provisioning eco-system services, and either in parallel or in succession (reuse) (Gordon *et al.*, 2010).

Recent basin-scale studies have demonstrated that by contrasting bright spots and hot spots, integrated water productivity assessments – bringing together crops, livestock, trees and fish – are useful means to identify tailored interventions (Ahmad and Giordano, 2010; van Breugel *et al.*, 2010; Cai *et al.*, 2011). At field level, crops with high water consumption such as rice can still be part of water-productive systems if their multiple agricultural (e.g. crop residues for feed), ecosystem (e.g. water flow regulation) and health (e.g. nutrition) services are taken into consideration (Matsuno *et al.*, 2002; Boisvert and Chang, 2006; Nguyen-Khoa and Smith, 2008). Hence, agricultural water management needs to focus on strategies that reduce costs, while at the same time aiming for greater integration between food production systems (such as crops, trees, livestock, aquaculture and fisheries), as well as safeguarding ecosystem services (Gordon *et al.*, 2010) (see Chapters 5 and 9). More water productivity gains could be made if not only food production systems, but the entire value chain, including postharvest losses, is considered (see Box 8.4).

Conclusions

Increasing the water productivity of crop, livestock and aquatic food production, while reducing social inequities and preserving the functioning of water bodies in a context of

> **Box 8.4.** Reducing postharvest losses.
>
> Approximately 1.3 billion t of food are lost or wasted annually, which is roughly one third of the human food produced (Gustavsson *et al.*, 2011). These losses occur mostly at the postharvest and processing levels in developing countries, and at the retail and consumer levels in industrialized countries (Gustavsson *et al.*, 2011). However, the per capita food losses in developing and industrialized countries are remarkably comparable. In sub-Saharan Africa, postharvest grain losses can amount to 10–20% of the production (World Bank *et al.*, 2011), which means that 10–20% of the inputs, including water, are wasted (Lundqvist *et al.*, 2008) as well. Therefore, reducing postharvest losses could be an effective way of achieving higher productivity (including water productivity) in agriculture (Clarke, 2004; INP*h*O, 2007). Many promising practices and technologies are available for reducing postharvest losses, including improved handling, storage and pest control (World Bank *et al.*, 2011). Incentives and public programmes are also needed to raise awareness and promote societal change in behaviour towards both a healthy diet and food waste.

increased demand for food and energy, is a real challenge. Consideration of the various ecosystem functions of irrigated and rainfed agroecosystems is essential, as is effective water governance at different scales, and attention to gender issues to help ensure sustainable and equitable use of water resources. In this chapter, the various options and solutions that are available for increasing agricultural water productivity have been reviewed. It has been demonstrated that going beyond crops, and including livestock, trees and fish in water productivity assessments, is crucial, and that many potential solutions are available. Greater awareness of these options among producers and policy makers can encourage more cost-effective water management strategies that can free up water for other uses, including ecosystem functioning.

An analysis of the effects of different options on future water demands from agriculture can be done through scenario analysis (e.g. de Fraiture and Wichelns, 2010). The inclusion of other sectors, such as livestock, fisheries, aquaculture and trees – as well as non-provisioning ecosystem services, makes it possible for such scenario analyses to contribute to a better understanding of the trade-offs between food, environment and the equitable distribution of gains (Cai *et al.*, 2011). Advances in modelling capabilities also enable impact assessments of climate change on the various components of agricultural water productivity. In addition, further research is needed on the implications of various (integrated) interventions and of improved agricultural water productivity on poverty, food security, economic growth and landscape functioning.

References

Adams, W.M., Watson E.E. and Mutiso, S.K. (1997) Water, rules and gender: water rights in an indigenous irrigation system, Marakwett, Kenya. *Development and Change* 28, 707–730.

Ahmad, M. and Giordano, M. (2010) The Karkheh River basin: the food basket of Iran under pressure. *Water International* 35, 522–544.

Al-Jamal, M.S., Ball, S. and Sammis, T.W. (2001) Comparison of sprinkler, trickle and furrow irrigation efficiencies for onion production. *Agricultural Water Management* 46, 253–266.

Amarasinghe, U., Palanisami, K. and Singh, O.P. (2012) Improving canal irrigation performance with on-farm water storage: evidence from the Indira Gandhi Nahar Pariyojana Project in India. *Irrigation and Drainage* 61, 427–435.

Amede, T., Descheemaeker, K., Peden, D. and van Rooyen, A. (2009a) Harnessing benefits from improved livestock water productivity in crop–livestock systems of sub-Saharan Africa: synthesis. *The Rangeland Journal* 31, 169–178. doi:10.1071/RJ09023

Amede, T., Geheb, K. and Douthwaite, B. (2009b) Enabling the uptake of livestock–water productivity interventions in the crop–livestock systems of sub-Saharan Africa. *The Rangeland Journal* 31, 223–230. doi:10.1071/RJ09008

Andrew, N., Béné, C., Hall, S.J., Allison, E.H., Heck, S. and Ratner, B.D. (2007) Diagnosis and management of small-scale fisheries in developing countries. *Fish and Fisheries* 8, 227–240.

Arora, V.K., Singh C.B., Sidhu, A.S. and Thind, S.S. (2011) Irrigation, tillage and mulching effects on soybean yield and water productivity in relation to soil texture. *Agricultural Water Management* 98, 563–568.

Balwinder, S., Humphreys, E., Eberbach, P.L., Katupitiya, A., Yadvinder, S. and Kukal, S.S. (2011) Growth, yield and water productivity of zero till wheat as affected by rice straw mulch and irrigation schedule. *Field Crops Research* 121, 209–225.

Barrios, E., Sileshi, G.W., Shepherd, K. and Sinclair, F. (2012) Agroforestry and soil health: linking trees, soil biota and ecosystem services. In: Wall, D.H., Bardgett, R.D., Behan-Pelletier, V., Herrick, J.E., Jones, H., Ritz, K., Six, J., Strong, D.R. and van der Putten, W.H. (eds) *Soil Ecology and Ecosystem Services*. Oxford University Press, Oxford, UK, pp. 315–330.

Bayala, J., Heng, L.K., van Noordwijk, M. and Ouedraogo, S.J. (2008) Hydraulic lift study in two native tree species of agroforestry parklands of West African dry savanna. *Acta Oecologia* 34, 370–378. doi:10.1016/j.actao.2008.06.010

Béné, C., Hersoug, B. and Allison, E.H. (2010) Not by rent alone: analyzing the pro-poor functions of small-scale fisheries in developing countries. *Development Policy Review* 28, 325–358.

Berejena, E., Ellis-Jones, J. and Hasnip, N. (1999) *Gender Sensitive Irrigation Design (Part 5): An Assessment of the Implications of Pump Breakdown and Community Participation in Masvingo Province, Zimbabwe.* Report OD 143, HR Wallingford, Wallingford, UK.

Berndes, G. (2002) Bioenergy and water – the implications of large-scale bioenergy production for water use and supply. *Global Environmental Change* 12, 253–271. doi:10.1016/S0959-3780(02)00040-7

Bindraban, P.S., Verhagen, A., Uithol, P.W.J. and Henstra, P. (1999) *A Land Quality Indicator for Sustainable Land Management: The Yield Gap.* Report 106, Research Institute for Agrobiology and Soil Fertility (AB-DLO), Wageningen, the Netherlands.

Bindraban, P.S., Bulte, E. and Conijn, S. (2009) Can biofuels be sustainable by 2020? *Agricultural Systems* 101, 197–199. doi:10.1016/j.agsy.2009.06.005

Bindraban, P.[S.], Conijn, S., Jongschaap, R., Qi, J., Hanjra, M., Kijne, J., Steduto, P., Udo, H., Oweis, T. and de Boer, I. (eds) (2010) *Enhancing Use of Rainwater for Meat Production on Grasslands: An Ecological Opportunity Towards Food Security. Proceedings 686, International Fertiliser Society, Cambridge, 10 December 2010, Leek, UK.* ISRIC World Soil Information, Wageningen, the Netherlands.

Boelee, E., Laamrani, H. and van der Hoek, W. (2007) Multiple use of irrigation water for improved health in dry regions of Africa and South-Asia. *Irrigation and Drainage* 56, 43–51. doi:10.1002/ird.287

Boisvert, R.N. and Chang, H.H. (2006) *Multifunctional Agricultural Policy, Reduced Domestic Support, and Liberalized Trade: An Empirical Assessment for Taiwanese Rice.* Comprehensive Assessment of Water Management in Agriculture Research Report 14, International Water Management Institute (IWMI), Colombo. doi:10.3910/2009.379

Bonachela, S., Orgaz, F. and Fereres, E. (1995) Winter cereals grown for grain and for the dual purpose of forage plus grain II. Water use and water-use efficiency. *Field Crops Research* 44, 13–24. doi:10.1016/0378-4290(95)00046-3

Bouma, J., Droogers, P., Sonneveld, M.P.W., Ritsema, C.J., Hunink, J.E., Immerzeel, W.W. and Kauffman, S. (2011) Hydropedological insights when considering catchment classification. *Hydrology and Earth System Sciences* 15, 1909–1919.

Bouman, B. (2007) A conceptual framework for the improvement of crop water productivity at different spatial scales. *Agricultural Systems* 93, 43–60.

Brummett, R.E., Lemoalle, J. and Beveridge, M.C.M. (2010) Can water productivity metrics guide allocation of freshwater to inland fisheries? *Knowledge and Management of Aquatic Ecosystems* 399, 1–7. doi:10.1051/kmae/2010026

Bunting, S.W. (2007) Confronting the realities of wastewater aquaculture in peri-urban Kolkata with bioeconomic modelling. *Water Research* 41, 499–505. doi:10.1016/j.watres.2006.10.006

Bunting, S.W. (2013) *Principles of Sustainable Aquaculture: Promoting Social, Economic and Environmental Resilience.* Earthscan, London, from Routledge, Oxford, UK.

Cai, X., Molden, D., Mainuddin, M., Sharma, B., Ahmad, M. and Karimi, P. (2011) Producing more food with less water in a changing world: assessment of water productivity in 10 major river basins. *Water International* 36, 42–62. doi:10.1080/02508060.2011.542403

Calzadilla, A., Rehdanz, K. and Tol, R.S.J. (2008) *Water Scarcity and the Impact of Improved Irrigation Management: A CGE Analysis.* Kiel Working Paper No. 1436, Kiel Institute for the World Economy, Kiel, Germany.

Capra, A., Consoli, S., Russo, A. and Scicolone, B. (2008) Integrated agro-economic approach to deficit irrigation on lettuce crops in Sicily (Italy). *Journal of Irrigation and Drainage Engineering* 134, 437–445.

Carroll, Z.L., Bird, S.B., Emmett, B.A., Reynolds, B. and Sinclair, F.L. (2004) Can tree shelterbelts on agricultural land reduce flood risk? *Soil Use and Management* 20, 357–359. doi:10.1111/j.1475-2743.2004.tb00381.x

Cecchi, G., Wint, W., Shaw, A., Marletta, A., Mattioli, R. and Robinson, T. (2010) Geographic distribution and environmental characterization of livestock production systems in Eastern Africa. *Agriculture, Ecosystems and Environment* 135, 98–110. doi:10.1016/j.agee.2009.08.011

Chapagain, A. and Hoekstra, A. (2003) Virtual water trade: a quantification of virtual water flows between nations in relation to international trade of livestock and livestock products. In Hoekstra, A.Y. (ed.) *Virtual Water Trade. Proceedings of the International Expert Meeting on Virtual Water Trade, IHE Delft, 12–13 December 2002.* Value of Water Research Report Series 12, United Nations Educational, Scientific and Cultural Organization Institute for Water Education (UNESCO-IHE), Delft, the Netherlands, pp. 49–76. Available at: http://www.unesco-ihe.org/content/download/1931/20430/file/Report12-hoekstra.pdf (accessed February 2013).

Clarke, B. (2004) *High-hopes for Post-harvest. A New Look at Village-scale Crop Processing.* FAO Diversification Booklet 4, Agricultural Support Systems Division, Food and Agriculture Organization of the United Nations, Rome.

Clement, F., Haileslassie, A., Ishaq, S., Blummel, M., Murty, M., Samad, M., Dey, S., Das, H. and Khan, M.A. (2011) Enhancing water productivity for poverty alleviation: role of capitals and institutions in the Ganga Basin. *Experimental Agriculture* 47(Supplement 1), 133–151. doi:10.1017/S0014479710000827

Cook, S.E., Andersson, M.S. and Fisher, M.J. (2009a) Assessing the importance of livestock water use in basins. *The Rangeland Journal* 31, 195–205. doi:10.1071/RJ09007

Cook, S.E., Fisher, M.J., Andersson, M.S., Rubiano, J. and Giordano, M. (2009b) Water, food and livelihoods in river basins. *Water International* 34, 13–29.

Cooper, P.J.M., Leakey, R.R.B., Rao, M.R. and Reynolds, L. (1996) Agroforestry and the mitigation of land degradation in the humid and sub-humid tropics. *Experimental Agriculture* 32, 235–290. doi:10.1017/S0014479700026223

de Fraiture, C. and Wichelns, D. (2010) Satisfying future demands for agriculture. *Agricultural Water Management* 97, 502–511. doi:10.1016/j.agwat.2009.08.008

de Fraiture, C., Giordano, M. and Yongsong, L. (2008) Biofuels and implications for agricultural water use: blue impacts of green energy. *Water Policy* 10 (Supplement 1), 67–81. doi:10.2166/wp.2008.054

de Vries, M.E., Rodenburg, J., Bado, B.V., Sow, A., Leffenaar, P. and Giller, K.E. (2010) Rice production with less irrigation water is possible in a Sahelian environment. *Field Crops Research* 116, 154–164. doi:0.1016/j.fcr.2009.12.006

Descheemaeker, K., Amede, T. and Haileslassie, A. (2010a) Improving water productivity in mixed crop–livestock farming systems of sub-Saharan Africa. *Agricultural Water Management* 97, 579–586. doi:10.1016/j.agwat.2009.11.012

Descheemaeker, K., Mapedza, E., Amede, T. and Ayalneh, W. (2010b) Effects of integrated watershed management on livestock water productivity in water scarce areas in Ethiopia. *Physics and Chemistry of the Earth* 35, 723–729. doi:10.1016/j.pce.2010.06.006

Descheemaeker, K., Amede, T., Haileslassie, A. and Bossio, D. (2011) Analysis of gaps and possible interventions for improving water productivity in crop–livestock systems of Ethiopia. *Experimental Agriculture* 47(Supplement 1), 21–38.

Dugan, P., Sugunan, V.V., Welcomme, R.L., Béné, C., Brummett, R.E., Beveridge, M.C.M. *et al.* (2007) Inland fisheries and aquaculture. In: Molden, D. (ed.) *Water for Food, Water for Life: Comprehensive Assessment of Water Management in Agriculture.* Earthscan, London, in association with International Water Management Institute (IWMI), Colombo, pp. 459–483.

Dugan, P. *et al.* (2010) Fish migration, dams, and loss of ecosystem services in the Mekong Basin. *Ambio* 39, 344–348.

Falkenmark, M. and Molden, D. (2008) Wake up to realities of river basin closure. *International Journal of Water Resources Development* 24, 201–215. doi:10.1080/07900620701723570

FAO (2008) *The State of Food and Agriculture 2008. Biofuels: Prospects, Risks and Opportunities.* Food and Agriculture Organization of the United Nations, Rome. Available at ftp://ftp.fao.org/docrep/fao/011/i0100e/i0100e.pdf (accessed December 2012).

FAO (2009) *Global Agriculture Towards 2050. How to Feed the World in 2050: High Level Expert Forum, Rome 12–13 October 2009.* Issues Paper HLEF2050, Food and Agriculture Organization of the United Nations, Rome. Available at: http://www.fao.org/fileadmin/templates/wsfs/docs/Issues_papers/HLEF2050_Global_Agriculture.pdf (accessed February 2013).

Ferraris, R. and Sinclair, D.F. (1980) Factors affecting the growth of *Pennisetum purpureum* in the wet tropics. II. Uninterrupted growth. *Australian Journal of Agricultural Research* 31, 915–925. doi:10.1071/AR9800915

Fischer, G., van Velthuizen, H., Hizsnyik, E. and Wiberg, D. (2009) *Potentially Obtainable Yields in the Semi-arid Tropics.* Global Theme on Agroecosystems Report No. 54, International Crops Research Institute for the Semi-Arid Tropics, Andhra Pradesh, India. Available at: http://www.iwmi.cgiar.org/assessment/files_new/publications/icrisatreport_54.pdf (accessed February 2013).

Gebreselassie, S., Peden, D., Haileslassie, A. and Mpairwe, D. (2009) Factors affecting livestock water productivity: animal scale analysis using previous cattle feeding trials in Ethiopia. *The Rangeland Journal* 31, 251–258. doi:10.1071/RJ09011

Geerts, S. and Raes, D. (2009) Deficit irrigation as an on-farm strategy to maximize crop water productivity in dry areas. *Agricultural Water Management* 96, 1275–1284. doi:10.1016/j.agwat.2009.04.009

Godfray, H.C.J., Beddington, J.R., Crute, I.R., Haddad, L., Lawrence, D., Muir, J.F., Pretty, J., Robinson, S., Thomas, S.M. and Toulmin, C. (2010) Food security: the challenge of feeding 9 billion people. *Science* 327, 812–818.

Goel A.K. and Kumar R. (2005) Economic analysis of water harvesting in a mountainous watershed in India. *Agricultural Water Management* 71, 257–266.

Gordon, L.J., Finlayson, C.M. and Falkenmark, M. (2010) Managing water in agriculture for food production and other ecosystem services. *Agricultural Water Management* 94, 512–519. doi:10.1016/j.agwat.2009.03.017

Gowda, C.L.L. *et al.* (2009) Opportunities for improving crop water productivity through genetic enhancement of dryland crops. In: Wani, S.P., Rockström, J. and Oweis, T. (eds) *Rainfed Agriculture: Unlocking the Potential.* Comprehensive Assessment of Water Management in Agriculture Series 7. CAB International, Wallingford, UK in association with International Crops Research Institute for the Semi-Arid Tropics (ICRISAT), Patancheru, Andhra Pradesh, India and International Water Management Institute (IWMI) Colombo, pp. 133–163.

Gustavsson, J., Cederberg, C., Sonesson, U., van Otterdijk, R. and Meybeck, A. (2011) Global *Food Losses and Food Waste.* Food and Agriculture Organization of the United Nations, Rome.

Hagos, F., Makombe, G., Namara, R.E. and Awulachew, S.B. (2009) *Importance of Irrigated Agriculture to the Ethiopian Economy: Capturing the Direct Net Benefits of Irrigation.* IWMI Research Report 128, International Water Management Institute, Colombo. doi:10.3910/2009.317

Haileslassie, A., Peden, D., Gebreselassie, S., Amede, T. and Descheemaeker, K. (2009a) Livestock water productivity in mixed crop–livestock farming systems of the Blue Nile basin: assessing variability and prospects for improvement. *Agricultural Systems* 102, 33–40. doi:10.1016/j.agsy.2009.06.006

Haileslassie, A., Peden, D., Gebreselassie, S., Amede, T., Wagnew, A. and Taddesse, G. (2009b) Livestock water productivity in the Blue Nile basin: assessment of farm scale heterogeneity. *The Rangeland Journal* 31, 213–222. doi:10.1071/RJ09006

Hall, S.J., Delaporte, A., Phillips, M.J., Beveridge, M.C.M. and O'Keefe, M. (2011) *Blue Frontiers: Managing the Environmental Costs of Aquaculture.* WorldFish, Penang.

Hansson, L. (2006) Comparisons of infiltration capacities in different parklands and farming systems of semi-arid Burkina Faso. Graduate Thesis in Soil Science, Department of Forest Ecology, Swedish University of Agricultural Sciences, Umeå, Sweden.

Harvey, C.A., Medina, A., Sanchez, D.M., Vilchez, S., Hernandez, B., Saenz, J.C., Maes, J.M., Casanoves, F. and Sinclair, F.L. (2006) Patterns of animal diversity in different forms of tree cover in agricultural landscapes. *Ecological Applications* 16, 1986–1999. doi:10.1890/1051-0761(2006)016[1986:POADID]2.0.CO;2

Hellegers, P., Zilberman, D., Steduto, P. and McCornick, P. (2008) Interactions between water, energy, food and environment: evolving perspectives and policy issues. *Water Policy* 10, 1–10.

Herrero, M., Thornton, P.K., Gerber, P. and Reid, R.S. (2009) Livestock, livelihoods and the environment: understanding the trade-offs. *Current Opinion in Environmental Sustainability* 1, 111–120. doi:10.1016/j.cosust.2009.10.003

Herrero, M. *et al.* (2010) Smart investments in sustainable food production: revisiting mixed crop–livestock systems. *Science* 327, 822–825. doi:10.1126/science.1183725

INP*h*O (2007) Information Network on Post-harvest Operations. Available at: www.fao.org/inpho/ (accessed December 2012).

Johnston, R. and McCartney, M. (2010) *Inventory of Water Storage Types in the Blue Nile and Volta River Basins.* IWMI Working Paper 140, International Water Management Institute, Colombo. doi:10.5337/2010.214

Jose, S. (2009) Agroforestry for ecosystem services and environmental benefits: an overview. *Agroforestry Systems* 76, 1–10. doi:10.1007/s10457-009-9229-7

Kandiero, T. (2009) *Infrastructure Investment in Africa.* Development Research Brief 10, Development Research Department, African Development Bank, Tunis Belvedere, Tunisia.

Kijne, J.W., Barker, R., and Molden, D. (eds) (2003) *Water Productivity in Agriculture: Limits and Opportunities for Improvement.* Comprehensive Assessment of Water Management in Agriculture Series 1. CAB International, Wallingford, UK in association with International Water Management Institute (IWMI), Colombo.

Kruseman, G., Ruben, R. and Tesfay, G. (2006) Diversity and development domains in the Ethiopian highlands. *Agricultural Systems* 88, 75–91.

Lemoalle, J. (2008) *Water Productivity of Aquatic Systems. Final Report Project CP-PN34: Improved Fisheries Productivity and Management in Tropical Reservoirs.* Challenge Program on Water and Food and WorldFish, Penang.

Lundqvist, J., de Fraiture, C. and Molden, D. (2008) *Saving Water: From Field to Fork: Curbing Losses and Wastage in the Food Chain.* SIWI Policy Brief, Stockholm International Water Institute (SIWI), Stockholm.

Matsuno, Y., Ko, H.S., Tan, C.H., Barker, R. and Levine, G. (2002) *Accounting of Agricultural and Nonagricultural Impacts of Irrigation and Drainage Systems: A Study of Multifunctionality in Rice.* IWMI Working Paper 43, International Water Management Institute (IWMI), Colombo.

McCartney, M. and Smakhtin, V. (2010) *Water Storage in an Era of Climate Change: Addressing the Challenge of Increasing Rainfall Variability.* IWMI Blue Paper, International Water Management Institute, Colombo.

Meinzen-Dick, R. (1997) Valuing the multiple uses of water. In: Kay, M., Franks, T. and Smith, L. (eds) *Water: Economics, Management and Demand.* E. and F.N. Spon, London, pp. 50–58.

Mekonnen, S., Descheemaeker, K., Tolera, A. and Amede, T. (2011) Livestock water productivity in a water stressed environment in northern Ethiopia. *Experimental Agriculture* 47(Supplement 1), 85–98.

Merrey, D.J. and Gebreselassie, T. (2011) *Promoting Improved Rainwater and Land Management in the Blue Nile (Abay) Basin of Ethiopia.* NBDC Technical Report 1, International Livestock Research Institute (ILRI), Nairobi.

Mintesinot, B., Verplancke, H., Van Ranst, E. and Mitiku, H. (2004) Examining traditional irrigation methods, irrigation scheduling and alternate furrows irrigation on vertisols in northern Ethiopia. *Agricultural Water Management* 64, 17–27.

Molden, D. (ed.) (2007) *Water for Food, Water for Life: Comprehensive Assessment of Water Management in Agriculture.* Earthscan, London, in association with International Water Management Institute (IWMI), Colombo.

Molden, D., Oweis, T.Y., Steduto, P., Kijne, J.W., Hanjra, M.A., Bindraban, P.S. *et al.* (2007) Pathways for increasing agricultural water productivity. In Molden, D. (ed.) *Water for Food, Water for Life: Comprehensive Assessment of Water Management in Agriculture.* Earthscan, London, in association with International Water Management Institute (IWMI), Colombo, pp. 279–310.

Molden, D., Oweis, T., Steduto, P., Bindraban, P., Hanjra, M.A. and Kijne, J. (2010) Improving agricultural water productivity: between optimism and caution. *Agricultural Water Management* 97, 528–535. doi:10.1016/j.agwat.2009.03.023

Moll, H.A.J., Staal, S.J. and Ibrahim, M.N.M. (2007) Smallholder dairy production and markets: a comparison of production systems in Zambia, Kenya and Sri Lanka. *Agricultural Systems* 94, 593–603. doi:10.1016/j.agsy.2007.02.005

Molle, F. and Renwick, M. (2005) *Economics and Politics of Water Resources Development: Uda Walawe Irrigation Project, Sri Lanka.* IWMI Research Report 87, International Water Management Institute, Colombo, Sri Lanka. doi:10.3910/2009.087

Molle, F., Wester, P. and Hirsch, P. (2010) River basin closure: processes, implications and responses. *Agricultural Water Management* 97, 569–577. doi:10.1016/j.agwat.2009.01.004

Mueller, N.D., Gerber, J.S., Johnston, M., Ray, D.K., Ramankutty, N. and Foley, J.A. (2012). Closing yield gaps through nutrient and water management. *Nature* 490 (7419), 254–257.

Mukherji, A., Facon, T., Burke, J., de Fraiture, C., Faures, J.M., Fuleki, B., Giordano, M., Molden, D. and Shah, T. (2009) *Revitalizing Asia's Irrigation: to Sustainably Meet Tomorrow's Food Needs*. International Water Management Institute (IWMI), Colombo and Food and Agriculture Organization of the United Nations (FAO), Rome.

Muthuri, C.W., Ong, C.K., Craigon, J., Mati, B.M., Ngumi, V.W. and Black, C.R. (2009) Gas exchange and water use efficiency of trees and maize in agroforestry systems in semi-arid Kenya. *Agriculture, Ecosystems and Environment* 129, 497–507. doi:10.1016/j.agee.2008.11.001

Mzezewa, J., Gwata, E.T. and van Rensburg, L.D. (2011) Yield and seasonal water productivity of sunflower as affected by tillage and cropping systems under dryland conditions in the Limpopo Province of South Africa. *Agricultural Water Management* 98, 1641–1648.

Nair, P.K.R., Kumar, B.M. and Nair, V.D. (2009) Agroforestry as a strategy for carbon sequestration. *Journal of Plant Nutrition and Soil Science* 172, 10–23. doi:10.1002/jpln.200800030

Nguyen-Khoa, S. and Smith, L.E.D. (2008) Fishing in the paddy fields of monsoon developing countries: re-focusing the current discourse on the 'multifunctionality of agriculture'. Keynote paper presented at the INWEPF-ICID Workshop of the RAMSAR COP10 Meeting, Changwon, Korea, 28 October–04 November 2008.

Nguyen-Khoa, S., Smith, L. and Lorenzen, K. (2005) *Impacts of Irrigation on Inland Fisheries: Appraisals in Laos and Sri Lanka*. Comprehensive Assessment Research Report 7, Comprehensive Assessment Secretariat, Colombo.

Nguyen-Khoa, S., van Brakel, M.L. and Beveridge, M.C.M. (2008) Is water productivity relevant in fisheries and aquaculture? In: Humphreys, E. *et al.* (eds) *Fighting Poverty Through Sustainable Water Use: Proceedings of the CGIAR Challenge Program on Water and Food 2nd International Forum on Water and Food, Addis Ababa, Ethiopia, November 10–14, 2008. Volume 1*. CGIAR Challenge Program on Water and Food, Colombo, Sri Lanka, pp. 22–27.

Nielsen, D.C., Vigil, M.F. and Benjamin, J.G. (2006) Forage yield response to water use for dryland corn, millet, and triticale in the Central Great Plains. *Agronomy Journal* 98, 992–998. doi:10.2134/agronj2005.0356

Nistor, A.P. and Lowenberg-DeBoer, J. (2007) Drainage water management impact on farm profitability. *Journal of Soil and Water Conservation* 62, 443–446.

Ong, C.K. and Leakey, R.R.B. (1999) Why tree–crop interactions in agroforestry appear at odds with tree–grass interactions in tropical savannahs. *Agroforestry Systems* 45, 109–129. doi:10.1023/A:1006243032538

Ong, C.K. and Swallow, B.M. (2003) Water productivity in forestry and agroforestry. In: Kijne, J.W., Barker, R. and Molden, D. (eds) *Water Productivity in Agriculture: Limits and Opportunities for Improvement*. Comprehensive Assessment of Water Management in Agriculture Series 1. CAB International, Wallingford, UK in association with International Water Management Institute, Colombo, pp. 217–228.

Ong, C.K., Anyango S., Muthuri, C.W. and Black, C.R. (2007) Water use and water productivity of agroforestry systems in semi-arid tropics. *Annals of Arid Zone* 46, 1–30.

Oweis, T. and Hachum, A. (2009a) Water harvesting for improved rainfed agriculture in the dry environments. In: Wani, S.P., Rockström, J. and Oweis, T. (eds) *Rainfed Agriculture: Unlocking the Potential*. Comprehensive Assessment of Water Management in Agriculture Series 7. CAB International, Wallingford, UK in association with International Crops Research Institute for the Semi-Arid Tropics (ICRISAT), Patancheru, Andhra Pradesh, India and International Water Management Institute (IWMI) Colombo, pp. 164–181.

Oweis, T. and Hachum, A. (2009b) Supplemental irrigation for improved rainfed agriculture in WANA region. In: Wani, S.P., Rockström, J. and Oweis, T. (eds) *Rainfed Agriculture: Unlocking the Potential*. Comprehensive Assessment of Water Management in Agriculture Series 7. CAB International, Wallingford, UK in association with International Crops Research Institute for the Semi-Arid Tropics (ICRISAT), Patancheru, Andhra Pradesh, India and International Water Management Institute (IWMI) Colombo, pp.182–196.

Oweis, T., Hachum, A. and Pala, M. (2004) Water use efficiency of winter-sown chickpea under supplemental irrigation in a Mediterranean environment. *Agricultural Water Management* 66, 163–179. doi:10.1016/j.agwat.2003.10.006

Peden, D., Ayalneh, W., El Wakeel, A., Fadlalla, B., Elzaki, R., Faki, H., Mati, B., Sonder, K. and Workalemahu, A. (2005) *Investment Options for Integrated Water-Livestock-Crop Production in Sub-Saharan Africa.* International Livestock Research Institute (ILRI), Addis Ababa.

Peden, D., Tadesse, G. and Misra, A.K. *et al.* (2007) Water and livestock for human development. In: Molden, D. (ed.) *Water for Food, Water for Life: Comprehensive Assessment of Water Management in Agriculture.* Earthscan, London, in association with International Water Management Institute (IWMI), Colombo, pp. 485–514.

Peden, D., Alemayehu, M., Amede, T., Awulachew, S.B., Faki, H., Haileslassie, A., Herero, M., Mapezda, E., Mpairwe, D., Musa, M.T., Taddesse, G. and van Breugel, P. (2009a) *Nile Basin Livestock Water Productivity, Project Number 37.* CPWF Project Report, CGIAR Challenge Program on Water and Food, Colombo.

Peden, D., Taddesse, G. and Haileslassie, A. (2009b) Livestock water productivity: implications for sub-Saharan Africa. *The Rangeland Journal* 31, 187–193. doi:10.1071/RJ09002

Pert, P.L., Butler, J.R.A., Brodie, J.E., Bruce, C., Honzák, M., Kroon, F.J., Metcalfe, D., Mitchell, D. and Wong, G. (2010) A catchment-based approach to mapping hydrological ecosystem services using riparian habitat: a case study from the wet tropics, Australia. *Ecological Complexity* 7, 378–388. doi:10.1016/j.ecocom.2010.05.002

Renault, D. and Wallender, W. (2000) Nutritional water productivity and diets. *Agricultural Water Management* 45, 275–296. doi:10.1016/S0378-3774(99)00107-9

Rockström, J. and Barron, J. (2007) Water productivity in rainfed systems: overview of challenges and analysis of opportunities in water scarcity prone savannahs. *Irrigation Science* 25, 299–311.

Rockström, J., Karlberg, L., Wani, S.P., Barron, J., Hatibu, N., Oweis, T., Bruggeman, A., Farahani, J. and Qiang, Z. (2010) Managing water in rainfed agriculture: the need for a paradigm shift. *Agricultural Water Management* 97, 543–550. doi:10.1016/j.agwat.2009.09.009

Roupsard, O. (1997) Ecophysiologie et diversité génétique de *Faidherbia albida* (Del.) A. Chev. (syn. *Acacia albida* Del.), un arbre à usages multiples d'Afrique semi-aride. Fonctionnement hydrique et efficience d'utilisation de l'eau d'arbres adultes en parc agroforestier et de juvéniles en conditions semi-contrôlées [Ecophysiology and genetic diversity of *Faidherbia albida* (Del.) A. Chev. (syn. *Acacia albida* Del.), a multipurpose tree of semi-arid Africa. Hydrological performance and water use efficiency of adult trees in agroforestry parkland and of young trees under semi-controlled conditions]. Thesis, Université H. Poincaré de Nancy I, Nancy, France.

Saeed, I.A.M. and El-Nadi, A.H. (1997) Irrigation effects on the growth, yield and water use efficiency of alfalfa. *Irrigation Science* 17, 63–68. doi:10.1007/s002710050023

Saeed, I.A.M. and El-Nadi, A.H. (1998) Forage sorghum yield and water use efficiency under variable irrigation. *Irrigation Science* 18, 67–71. doi:10.1007/s002710050046

Sala, O.E., Parton, W.J., Joyce, L.A. and Lauenroth, W.K. (1988) Primary production of the central grasslands of the United States. *Ecology* 69, 40–45. doi:10.2307/1943158

Sanchez, P.A. (1995) Science in agroforestry. *Agroforestry Systems* 30, 5–55. doi:10.1007/BF00708912

Schroth, G. (1999) A review of belowground interactions in agroforestry, focussing on mechanisms and management options. *Agroforestry Systems* 43, 5–34.

Schroth, G. and Sinclair, F.L. (eds) (2003) *Trees, Crops and Soil Fertility: Concepts and Research Methods.* CAB International, Wallingford, UK.

Seckler, D., Molden, D. and Sakthivadivel, R. (2003) The concept of efficiency in water resources management and policy. In: Kijne, J.W., Barker, R. and Molden, D. (eds) *Water Productivity in Agriculture: Limits and Opportunities for Improvement.* Comprehensive Assessment of Water Management in Agriculture Series 1. CAB International, Wallingford, UK in association with International Water Management Institute (IWMI), Colombo, pp. 37–51.

Shah, T. (2009) *Taming the Anarchy: Groundwater Governance in South Asia.* Resources for the Future, Washington, DC and International Water Management Institute (IWMI), Colombo.

Singh, O., Sharma, A., Singh, R. and Shah, T. (2004) Virtual water trade in dairy economy. Irrigation water productivity in Gujarat. *Economic and Political Weekly* 39, 3492–3497.

Singh, R.P., Ong, C.K. and Saharan, N. (1989) Above and below-ground interactions in alley cropping in semiarid India. *Agroforestry Systems* 9, 259–274. doi:10.1007/BF00141088

Siriri, D., Ong, C.K., Wilson, J., Boffa, J.M. and Black, C.R. (2010) Tree species and pruning regime affect crop yield on bench terraces in SW Uganda. *Agroforestry Systems* 78, 65–77. doi:10.1007/s10457-009-9215-0

Smeal, D., O'Neill, M.K. and Arnold, R.N. (2005) Forage production of cool season pasture grasses as related to irrigation. *Agricultural Water Management* 76, 224–236. doi:10.1016/j.agwat.2005.01.014

Steduto, P., Hsiao, T.C. and Fereres, E. (2007) On the conservative behavior of biomass water productivity. *Irrigation Science* 25, 189–207. doi:10.1007/s00271-007-0064-1

Steinfeld, H., Gerber, P., Wassenaar, T., Castel, V., Rosales, M. and de Haan, C. (2006) *Livestock's Long Shadow: Environmental Issues and Options.* Food and Agriculture Organization of the United Nations, Rome. Available at: www.fao.org/docrep/010/a0701e/a0701e00.HTM (accessed December 2012).

Stuyt, L.C.P.M., van Bakel, P.J.T., van Dijk, W., de Groot, W.J.M., van Kleef, J., Noij, I.G.A.M., van der Schoot, J.R., van den Toorn, A. and Visschers, R. (2009) *Samengestelde, Peilgestuurde Drainage in Nederland.* Voortgangsrapport 1 [*Compound Gauge Controlled Drainage in the Netherlands. Progress Report 1*], Alterra Research Institute, Wageningen University, Wageningen, the Netherlands.

Tarawali, S., Herrero, M., Descheemaeker, K., Grings, E. and Blümmel, M. (2011) Pathways for sustainable development of mixed crop livestock systems: taking a livestock and pro-poor approach. *Livestock Science* 139, 11–21. doi:10.1016/j.livsci.2011.03.003

Thornton, P. and Herrero, M. (2001) Integrated crop–livestock simulation models for scenario analysis and impact assessment. *Agricultural Systems* 70, 581–602. doi:10.1016/S0308-521X(01)00060-9

UNEP (2007) *Global Environment Outlook. GEO-4, Environment for Development.* United Nations Environment Programme, Nairobi. Available at: www.unep.org/geo/geo4.asp (accessed December 2012).

UNEP (2010) *Clearing the Waters. A Focus on Water Quality Solutions.* United Nations Environment Programme, Nairobi.

van Breugel, P., Herrero, M., van de Steeg, J. and Peden, D. (2010) Livestock water use and productivity in the Nile Basin. *Ecosystems* 13, 205–221. doi:10.1007/s10021-009-9311-z

van Noordwijk, M. and Ong, C.K. (1996) Lateral resource flow and capture – the key to scaling up agroforestry results. *Agroforestry Forum* 7, 29–31.

Verdegem, M.C.J., Bosma, R.H. and Verreth, J.A.J. (2006) Reducing water use for animal production through aquaculture. *Water Resources Development* 22 (1), 101–113. doi:10.1080/07900620500405544

Wichelns, D. (2010) Virtual water: a helpful perspective, but not a sufficient policy criterion. *Water Resources Management* 24, 2203–2219. doi:10.1007/s11269-009-9547-6

World Bank (2008) *Investment in Agricultural Water for Poverty Reduction and Economic Growth in Sub-Saharan Africa. Synthesis Report.* A Collaborative Program of AfDB (African Development Bank), FAO (Food and Agriculture Organization of the United Nations), IFAD (International Fund for Agricultural Development), IWMI (International Water Management Institute) and the World Bank. World Bank, Washington, DC. Available at: www.fanrpan.org/documents/d00508/ (accessed December 2012).

World Bank (2009) *Minding the Stock: Bringing Public Policy to Bear on Livestock Sector Development.* Report No. 44110-GLB, World Bank, Agriculture and Rural Development Department, Washington, DC.

World Bank, NRI and FAO (2011) *Missing Food: The Case of Post-harvest Grain Losses in Sub-Saharan Africa.* World Bank, with Natural Resources Institute, Medway, UK and Food and Agriculture Organization of the United Nations, Rome. World Bank, Washington, DC.

WRI, UNDP, UNEP and World Bank (2008) *World Resources 2008: Roots of Resilience: Growing the Wealth of the Poor – Ownership – Capacity – Connection.* World Resources Institute in collaboration with United Nations Development Programme, United Nations Environment Programme and World Bank. World Resources Institute, Washington, DC.

WWAP (2009) *The United Nations World Water Development Report 3 (WWDR3). Water in a Changing World.* World Water Assessment Programme, United Nations Educational, Scientific and Cultural Organization (UNESCO), Paris and Earthscan, London. Available at: http://unesdoc.unesco.org/images/0018/001819/181993e.pdf#page=5 (accessed February 2013).

Zomer, R.J., Trabucco, A., Coe, R. and Place, F. (2009) *Trees on Farm: Analysis of Global Extent and Geographical Patterns of Agroforestry.* ICRAF Working Paper No. 89, World Agroforestry Centre (ICRAF), Nairobi. Available at: www.worldagroforestrycentre.org/downloads/publications/PDFs/WP16263.PDF (accessed June 2011).

Zwarteveen, M.Z. (1995) *Linking Women to the Main Canal: Gender and Irrigation Management.* IIED Gatekeeper Series 54, International Institute for Environment and Development (IIED), London.

9 Managing Agroecosystem Services

**Devra I. Jarvis,[1*] Elizabeth Khaka,[2†] Petina L. Pert,[3]
Lamourdia Thiombiano[4] and Eline Boelee[5]**

[1]*Bioversity International, Rome, Italy;* [2]*United Nations Environment Programme
(UNEP), Nairobi, Kenya;* [3]*Commonwealth Scientific and Industrial Research
Organisation (CSIRO), Cairns, Queensland, Australia;* [4]*Central Africa Bureau, Food
and Agriculture Organization of the United Nations (FAO), Libreville, Gabon;* [5]
Water Health, Hollandsche Rading, the Netherlands

Abstract

Agriculture and ecosystem services are interrelated in various ways. Payments for ecological services (PES) and innovative methods of agricultural management, including ecological agriculture, conservation agriculture and the management of biological diversity are options for enhancing ecosystem services in agroecosystems while sustaining or increasing productivity. Successful actions will depend on strong supporting policies and legal frameworks, as well as on developing the knowledge and leadership capacity in farming communities to evaluate the potential benefits. The maintenance of ecosystem services and the long-term productivity and stability of agriculture ecosystems requires a paradigm shift in agriculture that moves away from single solutions to production problems towards a portfolio approach that supports multiple ways to better use soil, water and biotic resources to enhance ecosystem services.

Background

Agricultural production involves a wide range of ecosystem services and processes that use water, soil and biological components of the agricultural ecosystem, such as: nitrogen cycling, climate regulation, soil formation, pest and disease regulation and pollination, in addition to the obvious food production (Chapters 3 and 4). Some of these services are produced within the agricultural ecosystem itself while others rely on the supporting water, soil and biotic features of the environment that surround the agricultural production system. As weather patterns are becoming more unpredictable and extreme, with prolonged dry spells and very strong storm events (see Chapter 2), the concern over the long-term reduction in total water supply, and in the frequency and severity of pests and pathogens, calls for more attention to be given to the underlying ecosystem services that support these systems (Molden, 2007).

* E-mail: d.jarvis@cgiar.org
† E-mail: elizabeth.khaka@unep.org

In natural ecosystems, the relationship between diversity and ecosystem regulating and supporting services has been given economic value (Diaz and Cabido, 2001), but little attention has been focused on the ecological consequences of the loss of biotic diversity within agricultural ecosystems. This loss can affect the ecosystem regulating functions of agroecosystems, their capacity to support those ecosystem regulating services and the long-term stability of the ecosystem in the face of biotic and abiotic stresses (Hajjar *et al.*, 2008). In any ecosystem, each time a species or variety goes locally extinct, energy and nutrient pathways are lost, with consequent alterations of ecosystem efficiency and the ability of communities to respond to environmental fluctuations (Diaz and Cabido, 2001). Reduction of crop diversity, and of the associated diversity in agricultural landscapes, together with the associated reduction in functional traits and facilitative interactions, has reduced the capacity of agricultural ecosystems to regulate pests, diseases and pollinators, to recycle nutrients and to retain soil water (Hajjar *et al.*, 2008).

A fundamental research question emerges, therefore, on how to ensure that continued increases in agricultural intensification and productivity can be achieved in ways that use and enhance ecosystem services more effectively, as measured by increased stability and reduced variability in the agricultural production systems of small-scale farmers (Foley *et al.*, 2005; Tilman *et al.*, 2011). This includes increasing the adaptability of agricultural ecosystems in such a way that communities and agroecosystems are able to respond to changing conditions without debilitating losses in livelihoods, productivity or ecosystem functions.

As discussed in Chapter 4, ecosystem services in agriculture – that is, those other than the production of food or other agricultural products – have been assigned relatively low economic values compared with those in other natural ecosystems, largely as a result of a lack of understanding and limited data availability. However, 5 billion ha of land is currently cultivated or used for pasture. This is an area equal to approximately one third of the earth's total land area (Foley *et al.*, 2005), and it generates and interacts with an enormous range of agroecosystem services. There is a need to address this underestimation of ecosystem services in farmland, a need to develop concepts, policies and methods of evaluating them, and to find ways in which they can be maintained and enhanced in a way that is socially acceptable. Agroecosystems may very well offer the best chance of increasing global ecosystem services if land and water are managed in a way that enhances natural and social capital (Porter *et al.*, 2009). Specifically, enhancing the supporting and regulatory services of ecosystems is vital to meeting the food demands of a population forecast to reach 9 billion by 2050 (UNFPA, 2009).

Managing Ecosystem Services in Agriculture

Swinton *et al.* (2006) suggest that incentivizing a systems approach to agricultural management (rather than a problem-response approach) could support sustainable production as well as ecosystem services such as climate regulation, wildlife conservation, and biological pest control and pollinator management. Bennett *et al.* (2005) note that the ways in which ecosystems produce services are insufficiently understood, and that this uncertainty needs to be accounted for in the decision-making process. They advise that future management questions will have to address the complexity of ecosystems in their social context in order that ecological services can be maintained, and also to assess the degree to which technology can substitute for ecological services.

The ecosystem services framework provides a useful umbrella for this endeavour as it can only be achieved by healthy agroecosystems. Sustainable management plans have been advocated for various agroecosystems, ranging from hyper-arid and dryland systems (Chapter 6), to wetland and aquatic ecosystems (Chapter 7). Furthermore, as stated in Chapter 4, managing agroecosystems for the delivery of multiple services considerably improves the value of the land.

For instance, the on-site costs of nutrient depletion (including soil loss through erosion) in the agricultural sector of sub-Saharan Africa vary between countries from less than 1% to more than 20% of the agricultural gross domestic product (GDP) (Drechsel *et al.*, 2004). The off-site costs, especially in controlling erosion, can be much larger, and affect a variety of non-agricultural ecosystems and their services (Enters, 1998). The protection of these services by reducing soil, water and land degradation appears to be a cost-effective investment. Payments for environmental services (see below) and other finance mechanisms could be good incentives to use for stopping these off-site costs, but they would be context specific.

Managing livestock

With their many environmental impacts on soil, water and the atmosphere (Chapter 4), there are many opportunities for ecosystem gains in livestock production systems. For instance, the high emission of greenhouse gases can be mitigated by practices such as carbon sequestration in rangelands or improved pastures, by reversing deforestation for the production of feedstuffs through increased agricultural productivity and by using other methods of intensification (Watson *et al.*, 2000; Schuman *et al.*, 2002; Woomer *et al.*, 2004). Much can also be done by keeping fewer, but more productive, animals by means of better nutrition, animal health, breeding and husbandry techniques (Tarawali *et al.*, 2011). Another innovative approach is the establishment of community-based breeding programmes for the purpose of genetic improvement (e.g. in Ethiopia, where breeding animals are being selected based on phenotypes recorded within the village population; Mirkena *et al.*, 2011). To mitigate greenhouse gas emissions from animal waste, options lie in increased feed digestibility, better storage and treatment of the waste and the appropriate application of waste (World Bank, 2009). There are many other suggestions on how livestock could make a positive contribution to ecosystem services; however, the implementation of some of the proposed alternatives,

such as payments for carbon sequestration in rangelands, remains a challenge.

Similarly, health hazards and the pollution of land and water by livestock excreta could be turned around into enhanced nutrient cycling and increased soil water holding capacity by improved management practices. The most effective methods for addressing these problems in catchments are at the farm or production facility. Additional measures can control the effects of manure in watercourses, e.g. manure can be intercepted and stored in ponds, contaminated water can undergo on-farm treatment and constructed farm wetlands can be used to reduce the pathogen load (Dufour *et al.*, 2012). A potential method described by Masse *et al.* (2011) for developing more sustainable livestock operations utilizes anaerobic digestion biotechnologies to produce biogas, and by this means reduces the need for supplementary chemical nitrogen and phosphorus fertilizers.

The recovery of nutrients from manure, an important contribution to the supporting ecosystem service of nutrient cycling, is highly variable. Approximately 65% of manure nitrogen is recovered from (industrialized) intensive systems in Europe. Almost 30% of this is lost during storage and the maximum cycling efficiency as nitrogen available to crops is around 52%, with large differences between countries (Oenema *et al.*, 2007). In developing countries too, there is a large range of variation in nitrogen cycling efficiencies in manure management systems (Rufino *et al.*, 2006). Manure handling and storage, and synchronizing mineralization with crop uptake – and hence fine tuning nutrient cycling in the soil, are key ways in which nitrogen cycling efficiencies can be increased in mixed intensive systems, thus contributing to better regulation of water quality. Results from a recent study in England support earlier conclusions that additions of manure organic carbon produce measureable changes in a wide range of soil biophysical and physicochemical properties and processes that are central to the maintenance of soil fertility and functioning (Bhogal *et al.*, 2009, 2011). Smallholder farmers in Africa, who use little fertilizer, recognize the important role of manure in the efficient management and maintenance of soil fertility for crop production (Rufino *et al.*, 2007). Alternative management of livestock production systems shows that combinations of intensification, better integration of animal manure into crop production and matching the nitrogen and phosphorus supply to livestock requirements can effectively reduce nutrient flows (Bouwman *et al.*, 2011).

With respect to nutrient cycling, therefore, adjustments are needed both in nutrient-deficient systems, where soil fertility is being depleted, and in nutrient-loaded systems, where groundwater contamination, surface water eutrophication and soil pollution are major problems (World Bank, 2009). Technical solutions for reducing the quantity of animal waste and facilitating its proper management and application have to be supported by regulatory measures and financial instruments, such as subsidies and taxes. In nutrient-deficient systems, the proper integration of livestock and crop production components in mixed and agropastoral systems can alleviate nutrient export through the application of manure and urine to cultivated areas (Powell *et al.*, 2004).

Trees for agroecosystem services

A long tradition of separate science and practice in forestry and agriculture means that there are largely untapped opportunities for using trees constructively in agricultural landscapes to sustain food production, while improving a range of ecosystem services. Trees have great potential to play an important role in the sustainable management of agro-ecosystems. In addition to having impacts on the supporting, regulatory, and cultural services of ecosystems, trees in agroecological land-scapes may increase provisioning services by contributing fruit, fodder, fuelwood and timber.

The impact of changing tree cover on various ecosystem services depends on its amount, spatial configuration, species com-position and management. So there is a need to consider planned tree cover change at a landscape scale with the aim of meeting specific suites of objectives, including consideration of the trade-offs and synergies among the ecosystem services affected (Jackson *et al.*, 2013; see also Box 4.1, Chapter 4). The enhancement of tree cover on farmland has the potential to tighten nutrient, water and carbon cycles, and promote the abundance and activity of soil organisms (Barrios *et al.*, 2012), thereby increasing and sustaining soil and water productivity. Different tree species root to different depths, have leaves at different times throughout the year, and use more or less water through transpiration, attributes that are all affected by management practices such as pruning.

Land management

A variety of soil conservation techniques are available that can be integrated into agricultural and other land use practices to sustain and enhance agroecosystems and minimize their adverse impacts on their closer environment (Bindraban *et al.*, 2012). Integrated solutions for tackling land degradation can lead to improved water productivity and environmental health (Descheemaeker *et al.*, 2009), without reducing water availability for food and feed production. An example from Ethiopia describes how successful approaches integrate water and land management with improved agricultural practices (Box 9.1), but more examples exist of solutions developed for multifunctional agroecosystems (e.g. Matsuno *et al.*, 2002; Vereijken, 2003; Boody *et al.*, 2005; Boisvert and Chang, 2006; Nguyen-Khoa and Smith, 2008).

Payments for Ecosystem Services

Payments for ecosystem services (PES), also known as payments for environmental services (or benefits), is the practice of compensating individuals or communities for undertaking actions that increase the provision of ecosystem services such as water purification, flood mitigation and carbon sequestration (Kelsey Jack *et al.*, 2008). PES comes under the heading of economic or market-based incentives aimed at motivating the desired decision taking through charges, tradable permits, subsidies and market friction reductions. While the term 'PES' has been in

Box 9.1. Integrated watershed management for improved water productivity and ecosystem services in Ethiopia.

Crop–livestock farming is an important livelihood strategy for smallholder farmers in water-scarce areas of Ethiopia, which are characterized by land degradation, low agricultural productivity, food insecurity and increasing population pressure (Descheemaeker *et al.*, 2010b). Integrated watershed management has become a popular way to tackle the interrelated problems of land degradation, low productivity, institutional and organizational constraints and poverty (German *et al.*, 2007; Shiferaw *et al.*, 2009). Community-based integrated watershed management – through exclosures (areas closed for grazing and agriculture) and water-harvesting ponds – was implemented in the water-scarce Lenche Dima watershed in the northern highlands of Ethiopia (Liu *et al.*, 2008).

Exclosures were established on the degraded hill slopes in the watershed with the overall aim of rehabilitating the area (Descheemaeker *et al.*, 2010b). In these closed areas, contour trenches were established to improve water infiltration, and multipurpose trees were planted at the time of closing. These actions enhanced both regulatory (water regulation) and supporting (soil formation) ecosystem services. The community was responsible for the protection of the area and this was institutionalized through written by-laws. Provisioning services were also enhanced as the production of herbaceous and woody biomass in exclosures recovered dramatically (Fig. 9.1), and farmers harvested the grass for haymaking. The exclosures led to improvements in livestock water productivity as well (Descheemaeker *et al.*, 2009): by protecting about 40% of the rangelands in the watershed, the water productivity of the feed increased by 18–49%, depending on the amount of hay produced in the exclosures. As a result, the livestock production per unit of water depleted increased. Long-term environmental benefits (observed runoff reduction, groundwater recharge and the protection of downstream cropland from peak flows) and increased woody biomass production from the exclosures contributed to improved ecosystem services in the watershed (Descheemaeker *et al.*, 2010b).

Fig. 9.1. Degraded open access grazing land (left) and protected exclosures 3 years after closing (right) in Ethiopia (photos by Katrien Descheemaeker).

The second intervention was the construction of dome-shaped water harvesting structures in the farmers' homesteads (Descheemaeker *et al.*, 2010b). On average, farmers used 50% of the water to irrigate the fruit trees and vegetables planted in their homesteads. Domestic uses accounted for about 20% of the water use, and livestock drinking for the remaining 30%, mostly in the dry period. The effect of the water harvesting structures on livestock water productivity was brought about through the reduction of the energy spent by the animals in walking to the drinking points in the dry season (about 11% of their annual energy budget). This saved energy could, potentially, be used for productive purposes such as milk production (Descheemaeker *et al.*, 2010a). Other studies (Muli, 2000; Staal *et al.*, 2001; Puskur *et al.*, 2006) found that water harvesting structures enabled farmers to combine vegetable production with small-scale dairy farming, which significantly increased milk production and farmers' incomes. While animals were kept in the homestead for drinking, the pressure on the rangelands was reduced too, thus avoiding land degradation and the disruption of environmental flows (Descheemaeker *et al.*, 2010b).

common use since the 1990s, PES type schemes have been around since at least the 1930s when, in the wake of the American Dust Bowl, the federal government paid farmers to avoid farming on poor quality erodible land.

Various case studies are discussed in Dunn (2011); these look at the changing drivers for agriculture and at growing urbanization, which both threaten water quality, and at how organizations have set up PES schemes with local farmers. For example, companies pay farmers to adopt less intensive farming techniques, such as outdoor grazing, instead of fertilizer-intensive crop cultivation and feedlots, and the planting of trees to improve soil conditions and promote filtration services. Payments provide sufficient incentives to compensate the famers for these actions, and are developed in collaboration with famers and academics, and negotiated with each farmer. They are intended to reward services that go beyond what is legally required. Such schemes have documented successes in terms of their impacts on water quality, farmer

profitability and biodiversity outcomes (see Dunn, 2011).

In several of the integrated watershed programmes that have been implemented in India, upstream farmers are compensated for changing their practices, but not necessarily always in cash (Box 9.2). Hence, demand for a wide range of ecosystem services from agriculture will increase owing to a greater awareness of both their value and the costs inherent in their depletion (FAO, 2007).

Today, there are literally hundreds of ongoing PES schemes of all shapes and sizes, all over the world. Some are directed towards achieving poverty reduction on a local level; others maximize the output of goods on an industrial scale. However, all of the schemes essentially involve three steps (WWF, 2010). First, an assessment of the range of ecosystem services that flow from a particular area, and who they benefit. Secondly, an estimate of the economic value of these benefits to the different groups of people. Finally, a policy, a subsidy or a market to capture this value and compensate individuals or communities for

---

Box 9.2. Payments for water services in Sukhomajri, India.

The small village of Sukhomajri in the foothills of the Shivaliks provides an early and complex example of watershed development that has helped to inspire modern watershed development programmes (FAO, 2007). In the 1970s, high rates of sedimentation in Lake Sukhna in the northern Indian state of Haryana created problems for the drinking water supply of the nearby town of Chandigarh (Kerr, 2002). The source of the problem was traced to a small upstream village named Sukhomajri, where villagers were cultivating steep lands, and allowing animals to graze freely throughout the watershed. Around 80–90% of the sedimentation in Lake Sukhna was found to originate from Sukhomajri (Sengupta et al., 2003). The agricultural practices of the Sukhomajri farmers were not only felt downstream, but also in the village itself, where runoff water on one side of the watershed flooded and destroyed agricultural lands.

A central government agency, the Central Soil and Water Conservation Research and Training Institute (CSWCRTI) revegetated the watersheds and installed conservation structures such as check dams and gully plugs to stop the flow of silt. Villagers were asked to refrain from allowing grazing animals on to the watersheds. Benefits to the villagers were twofold: damage to agricultural lands was reduced, and there was access to irrigation water stored by the check dams. Although no direct payments were involved, the villagers were thus indirectly compensated for providing the environmental service. At the time of the implementation of the project, the notion of markets for environmental services was little known but, in effect, the project functioned as an environmental services payment scheme.

A drawback was that only a minority of landowners in the village benefited from the scheme; other villagers, particularly the landless, stood to lose from reduced access to grazing lands. The problem was solved by distributing rights to the water to all villagers and allowing them to trade among themselves – a system that was later abandoned in favour of user fees for water. The project resulted in a 95% decrease in siltation into Lake Sukhna, and saved the town of Chandigarh about US$200,000 annually (Kerr, 2002).

their action. In China's renowned 'Grain for Green' programme, the government thus compensates farmers with grain and cash for planting trees on their sloping farmlands (Box 9.3).

Developing mechanisms to implement PES is challenging, not least because although the concept is simple, the reality of making such schemes operational can be very complex, and budgetary resources are often a constraint – especially in poorer countries. Nevertheless, PES can trigger creativity in finding innovative solutions. When effectively designed, PES schemes can give both providers and users of ecosystem services more accurate indications of the consequences of their actions, so that the mix of services provided matches more closely the true preferences of the society concerned (FAO, 2007). This is the case in Brazil, where water users pay for measures that prevent pollution and erosion (Box 9.4). Water users themselves rarely take the initiative but, in Nepal, a fishing community has developed its own, demand-led, mechanism to ensure good water quality (Box 9.5).

A related and comparable concept is that of green water credits, where incentives are given for sound water management or sediment control by appropriate tillage methods or other eco-efficient farming techniques (Dent and Kauffman, 2007; Jansen et al., 2007). The idea is to create investment funds so that farmers can take intervention measures for better management of soil and water upstream, which will then be paid for by downstream users that receive more and better quality water.

Box 9.3. China's Grain for Green programme.

Pushed into action by a series of devastating floods in 1998, the Chinese government launched the Grain for Green programme in 1999 (FAO, 2007). This is one of the largest conservation set-aside programmes in the world, and its main objective is to increase forest cover on sloped cropland in the upper reaches of the Yangtze and Yellow River basins to prevent soil erosion. When possible in their community, households set aside all or parts of certain types of land and plant seedlings to grow trees. In return, the government compensates the participants with grain, cash payments and free seedlings. By the end of 2002, officials had expanded the programme to some 15 million farmers in more than 2000 counties in 25 provinces and municipalities (Xu et al., 2004). A recent impact analysis of 11 river basins covered by the Grain for Green programme suggests that both runoff and soil erosion have been reduced (Deng et al., 2012).

Box 9.4. Brazil's Water Producer Programme (TNC, 2008)

The Paraná River is the second longest river in South America, running through Brazil, Paraguay and Argentina over a course of 2570 km. The river provides multiple ecosystem services to the populations living within its watershed, including water for irrigation and the provision of drinking water to South America's largest city, São Paulo. However, the water quality of the Paraná River has declined over time as a result of the intensive deforestation of the Atlantic Forest at its headwaters. Without forest cover around the river's edge (the riparian zone), rainwater washes away soil, leading to a build-up of sediment that alters the water quality and may invade irrigation systems.

In an effort to improve the water quality of the Paraná River while at the same time protecting the biodiversity of the Atlantic Forest, The Nature Conservancy (an international organization) developed the Water Producer Programme, and it is implemented by Brazil's National Water Agency (ANA), the Agriculture and Environment Secretaries of São Paulo, the Piracicaba–Capivari–Jundiai (PCJ) watershed committee and the municipal government of Extrema in the state of Minas Gerais. The programme proposes using a portion of the water fees collected from major water users, such as water supply companies, and major industries to plant trees along riparian zones in the river's headwaters. These activities are executed by farmers and ranchers who receive a payment to reforest and maintain key sections of their land that are critical to the health of the Paraná River, thus contributing to the regulatory services of the river. Landowners also receive technical assistance on reforestation, soil conservation and erosion prevention from the programme's partners.

Box 9.5. The Rupa Lake Cooperative, Nepal (Pradham *et al.*, 2010).

Rupa Lake is the third largest lake (area 1.35 km²) in Nepal. It is located in the mid-western part of the country at an altitude of about 600 m asl. The area was once rich in biodiversity, but the ecosystem had deteriorated over the last few decades because of human encroachment of the land around the lake. Its conversion to agriculture had resulted in an increase in heavy landslides, pollution by chemical waste and the silting of downstream areas, all of which threatened the livelihoods of the fishing households earning their living from the lake.

The Rupa Lake Restoration and Fishery Cooperative, founded in 2001 by a downstream community for which fishery is an important part of their livelihood strategy, established a benefit-sharing mechanism to provide incentives to communities and various upstream user groups to conserve the catchment. The process was developed through local, traditional mechanisms, in the absence of official markets for the environmental services. The Rupa Cooperative decided to pay 10% of its income from fishery management to the upstream communities with the aim of ensuring good upstream crop management practices to reduce siltation and promote water quality. The payment mechanism is voluntary, and there is no contract or agreement made between the buyers (the Cooperative) and the sellers (the upstream users). Direct payments are made by the Cooperative on an annual basis to different user groups, such as Community Forest User Groups, schools and communities who request funding for specific watershed management activities. Rewards or indirect payments are also made by the Cooperative in kind through the provision of seedlings and gabion boxes.

Ecological Agriculture

Another way of targeting more ecosystem services in agriculture is through alternative approaches to agriculture that are more sustainable and safeguard ecosystem services, in particular from the point of view of water management. Several tools and approaches have been used to implement the concept of sustainable agriculture, such as sustainable land management, ecoagriculture, conservation agriculture, conservation farming, organic agriculture, increased genetic diversity in the production system and others (Francis and Porter, 2011; Gomiero *et al.*, 2011; Mulumba *et al.*, 2012). There are also successful local experiences that have made a paradigm shift away from single solutions to using a portfolio of methods to promote sustainable agriculture; this process should meet the following criteria (FAO, 1995):

- Ensure that the basic nutritional requirements of present and future generations are met both qualitatively and quantitatively, while providing a number of other agricultural products and ecosystem services.
- Provide durable employment, sufficient income, and decent living and working conditions for all those engaged in agricultural production.
- Maintain and, where possible, enhance the productive capacity of the natural resource base as a whole, and the regenerative capacity of renewable resources, without disrupting the functioning of basic ecological cycles and natural balances, or destroying the sociocultural attributes of rural communities, or causing contamination of the environment.
- Reduce the vulnerability of the agricultural sector to adverse natural and socio-economic factors and other risks, and strengthen self-reliance.

Examples of such successful local experiences include two from Kenya: the programme to regain the eroded uplands of Machakos by the Akamba people (summarized in UNDP *et al.*, 2000); and projects carried out by SACDEP-Kenya (Sustainable Agriculture Community Development Programmes in Kenya) (outlined in Box 9.6).

Conservation agriculture also tries to increase ecosystem services in agriculture, mainly through reducing tillage and restoring land cover, as shown by an example from Zambia (Box 9.7). Its primary purpose is to bring water back into the soil and keep it there,

Box 9.6. Small-scale sustainable agriculture in Kenya.

Since 1993, SACDEP-Kenya (Sustainable Agriculture Community Development Programmes in Kenya; see http://sacdepkenya.org/) trained over 40,000 farmers in 14 districts in Kenya. During those years, the strategies of sustainable agriculture have been refined. While conventional agriculture is mainly about increased production and incomes, SACDEP uses four principles to guide sustainable agriculture: that it be economically feasible, environmentally friendly, socially just and culturally acceptable. In order to make these principles practically operational, the necessary pillars of sustainable agriculture were defined. These pillars are based on farmer working groups, low-cost external inputs, organic agriculture, the ability of communities to mobilize finances, renewable energy, farmers' participation in conservation, and processing and value addition; they also include marketing decisions (including pricing) and the formulation of policies for agricultural and rural development. SACDEP has had successful projects in Kenya on organic products, draft animal power, low cost livestock (such as dairy goats), wind energy, Direct Organic Markets and high value alternative and emerging crops. It would be interesting to measure the impact of the combined interventions on ecosystem services, particularly on regulatory and supporting services, such as ecosystem resilience.

Box 9.7. Conservation farming in Zambia.

As an example of local initiatives in Africa, the PELUM Association (www.pelumrd.org) is a network of 207 civil society organizations in eastern, central and southern Africa that is working towards poverty eradication and food security through sustainable agriculture. It aims to build the capacity of farming and rural community groups to accumulate skills, to stimulate farmer learning and to inspire experimentation and innovation in the quest to achieve food security. In doing this, it builds on the potential of indigenous knowledge and indigenous farming and cropping patterns.

A study by PELUM on 15 small farms and two commercial farms in Zambia before and after conversion towards conservation farming showed that it can be an important first step to enabling smallholder farmers to get out of poverty and towards sustainable farming:

- Conventional small-scale farming in Zambia had nationwide average yields of 1.1 t/ha, and mostly economic deficits, because of the high costs related to inputs such as tillage and fertilizer.
- Almost a third of all fields were abandoned at the time of harvest every year, because inputs (labour, ploughs, fertilizers) were not available at the right time.
- In the 'worst' sub-village, a pilot project with technical support from PELUM achieved a 70% increase of yield and profit after 6 days of training and individual coaching.
- A comparison between various ploughing techniques and implements showed that:
  - Ploughing led to the lowest yields (average 2.4 t/ha)
  - Ripping was better (yields about 4 t/ha)
  - Hand hoeing gave the best results (yields 5–8 t/ha)
  - The highest yields of 8 t/ha were only reached by farmers who used manure (chemical fertilizers showed lower yields).

The sustainability of farms was measured before and after the conversion to conservation farming. Profit was the indicator for economic sustainability, while for ecological sustainability carbon dioxide (CO_2) equivalents were used. In Zambia, conservation farming proved to be significantly more profitable (70% more profit 1–2 years after conversion) than conventional farming. This applied to small and large farms applying zero tillage and direct drilling into the stubble. Although ecosystem services were not explicitly measured by PELUM, it appears, in any case, that the supporting service of soil formation was enhanced.

but it can have much larger benefits, such as shown by the example of Itaipu, in Brazil (Box 9.8). The Itaipu case demonstrates that, by considering and managing ecosystem functions and services, win–win solutions for both agriculture and other needs can be achieved. The interventions made have increased agricultural productivity and sustainability, in addition to delivering benefits to other ecosystems, such as reduced erosion. The

Box 9.8. Conservation agriculture in the Itaipu watershed, Brazil.

Farming activities in the Itaipu watershed, in the Paraná Basin in Brazil, were a significant threat to the Itaipu dam, a major facility generating hydroelectric power for Brazil, Argentina and Paraguay. The promotion of conservation agriculture in this watershed has enabled farmers to deliver improved ecosystem services, in particular through the reduction of soil erosion and the delivery of clean water to the reservoir (Mello and van Raij, 2006; ITAIPU, 2011). Not only did this approach improve farmer livelihoods, it also extended the life expectancy of the dam fivefold. This translated into a considerable benefit, considering the original investment costs of the dam and its regional economic importance. Furthermore, as in many cases, irrespective of increased farm profitability, the on-farm value of agricultural produce (direct farm profits) was eclipsed by the value of the improved catchment services provided through more sustainable farming.

move from conventional agricultural and environmental management practices to non-conventional practices such as conservation agriculture represents a great challenge in terms of changing habits and minds (Table 9.1).

Ecoagriculture is another of the many approaches towards sustainable farming, and is highlighted in this book because of its landscape scale and its compatibility with modern high input agriculture (see also Chapter 11). It is 'the design, adaptation and management of agricultural landscapes to produce ecosystem services (e.g. watershed services, wild biodiversity) and generate positive co-benefits for production, biodiversity, and local people, while addressing climate change challenges' (Scherr and McNeely, 2008; Ecoagriculture Partners, 2012). Such integrated agricultural landscapes provide critical watershed functions through careful rain and soil water management. This integrated management encompasses the choice of water-conserving crop mixtures, soil and water management (including irrigation), the maintenance of soils to facilitate rainfall infiltration, vegetation barriers to slow the movement of water down slopes, year-round soil cover, and maintenance of natural vegetation in riparian sites, wetlands and other strategic areas of the watershed.

Parallel to the demand for more sustainable agriculture, the health sector has developed interdisciplinary approaches such as 'One

Table 9.1. Comparison of conventional farming with conservation agriculture (from Thiombiano and Meshack, 2009).

Farming practice	Conventional farming	Conservation agriculture	Rationale
Tillage	Farmers plough and hoe to improve the soil structure and control weeds	Direct planting without prior inversion of the soil Planting on the rip line or making holes for planting with a hoe	In the long term, ploughing destroys the soil structure and contributes to declining fertility and levels of organic matter
Crop residues	Farmers remove or burn residues or mix them into the soil with plough or hoe	Crop residue left on the field Planting of cover crops	Crop residues improve soil structure Cover crops protect soil from erosion and limit weed growth
Mix and rotate crops	Monocultures or crop rotations in a tillage framework where the soil is inverted with a mouldboard plough or similar implement	Crop rotation or intercropping is a permanent feature of the cropping system	Helps to maintain soil fertility Breaks disease cycles

Health', striving to attain optimal health for people, animals and our environment, and 'Ecohealth', a participatory methodology for understanding and promoting health and well-being in the context of social and ecological interactions. Both of these methods fit well within an ecological approach to agriculture as the two integrated health approaches emphasize a multidisciplinary process and the importance of agriculture and ecosystem-based interventions (Waltner-Toews, 2009). This makes them highly suitable for addressing water-related diseases, in a manner that is complementary to that of sustainable agri-culture (see Chapter 5). Agricultural practices that create health risks, such as those related to water management, obviously require farm-level interventions, and food-borne diseases require management along the 'field-to-fork', or 'boat-to-throat' risk pathway. This includes management of water used at different stages, be it as a production input, in processing, or in meal preparation. Most zoonoses need veterinary and agroecological interventions in addition to medical interventions, as they cannot be controlled as long as diseases remain in the animal reservoir. For zoonoses transmitted through water (e.g. leptospirosis) or via aquatic hosts (e.g. schistosomiasis) interventions may also need to be directed at the aquatic ecosystems.

Managing Biological Diversity Within Agroecosystems

Recently, more attention has been given to the role of the biological diversity of cultivated ecosystems in providing ecosystem regulating and supporting services (FAO and PAR, 2011). There is a growing body of literature that functional diversity – the value and range of species traits rather than just species numbers – is important to short-term ecosystem resource dynamics and long-term ecosystem stability, as it increases positive interactions or complementary functions (Diaz and Cabido, 2001; Wilby and Thomas, 2007). First, crop genetic diversity has been shown to have a direct effect on the maintenance of ecosystem services by providing both: (i) increased numbers of functional traits; and (ii) facilitative

interactions that maintain above- and below-ground associated biodiversity. This has been shown to be useful in pest and disease management, and has the potential to enhance pollination services and soil processes (nutrient cycling, decomposition and erosion control) in specific situations (Hajjar *et al.*, 2008). Secondly, by increasing long-term stability of the ecosystem in the face of biotic and abiotic stresses and socio-economic variability, crop genetic diversity promotes the continuous maintenance of biomass and the ecosystem services that it provides.

Maintaining or increasing the genetic diversity within the farmer's production system through the use or development of varietal mixtures, or of sets of varieties with non-uniform resistance, has been an alternative agricultural management practice for regulating pests and diseases in many parts of the world (Finckh *et al.*, 2000; Finckh and Wolfe, 2006). The main purpose of genetic mixtures (crop variety mixtures) for pest and disease management is to slow down the spread of pests and pathogens (Wolfe, 1985). Recent studies have shown that a diverse genetic basis of resistance is beneficial for the farmer because it allows a more stable management of pest and disease pressure than does a monoculture (Trutmann *et al.*, 1993; Thurston *et al.*, 1999; Thinlay *et al.*, 2000; Finckh, 2003; Di Falco and Chavas, 2007; Jarvis *et al.*, 2007). The high levels of diversity of traditional rice varieties in Bhutan have been shown to have high functional diversity against rice blast (Thinlay *et al.*, 2000; Finckh, 2003). Increased levels of common bean and banana diversity in Uganda when disease levels were high showed a significant reduction in pest and disease damage in farmers' fields (Mulumba *et al.*, 2012; Box 9.9).

There is growing evidence of the potential of crop genetic diversity to enhance an agroecosystem's capacity to sustain biomass levels through improving the resilience and resistance to environmental variability of that system (Sadiki, 2006; Sawadogo *et al.*, 2006; Weltzien *et al.*, 2006). High levels of crop genetic diversity occur most commonly in areas where the production environment itself is extremely variable. Here, crop genetic diversity, through its increased portfolio of

---

Box 9.9. Crop varietal diversity to regulate pests and diseases in Uganda (Mulumba *et al.*, 2012).

Bananas and plantains (*Musa* spp.) and common beans (*Phaseolus vulgaris*) are important carbohydrate sources for local people in Uganda. Both crops are maintained as a mixture of different genotypes in farmers' fields. The varietal diversity of the local crops was measured at both community and household levels within 60 farmers' fields in each of four agroecological areas of Uganda. Participatory diagnostics of farmers' knowledge linked to cross-site, on-farm and on-station trials was then used to assess the resistance of traditional and modern varieties of *P. vulgaris* to anthracnose, angular leaf spot and bean fly, and of traditional and modern varieties of *Musa* spp. to black sigatoka, banana weevils and nematodes; the assessments of resistance were then compared with the intraspecific diversity of these two crops in the farmers' fields.

A general trend for both crops was that with increased diversity of crop varieties, as measured by the number of varieties (varietal richness) and their evenness of distribution, there was a decrease in the average damage levels across sites. Moreover, this increased diversity was related to a reduction in the variance of disease damage. That there was a reduction in the variance of disease damage as the diversity increased is an indication that some of the uniform farms (i.e. those growing a particular variety) will be fine, but only in the case that they happen to be growing a winning variety for that year; otherwise, these farms will be hit far worse in terms of crop damage when there is a change in pathogen or pest biotype.

The results support what might be expected in a risk-minimizing argument for using diversity to reduce pest and disease damage: diversity may both reduce current crop damage and have the potential to reduce future vulnerability to pest and disease infestations. The relationship of increased diversity to decreased damage was particularly evident when the damage of the disease was higher i.e. in sites with higher disease incidence, households with higher levels of diversity in their production systems had less damage to their standing crop in the field.

---

types, provides the capacity to cope with multiple stresses and changing conditions, thereby ensuring a more stable vegetation cover under a less predictable environment (Brush 1991; Aguirre *et al.*, 2000; Hajjar *et al.* 2008).

The provision of ecosystem services that support soil, water and nutrient availability (Chapter 3), and consequently biomass yield, is a management issue that also has the potential to be addressed through crop genetic diversity. In Nepal, farmers typically plant several varieties of rice to match the soil, moisture and other micro-ecological conditions in upland, lowland and swamp environments, which are often all found on the same farm. More than twice the number of rice varieties are found in the hills (which are generally more prone to erosion) as in the lowlands; moreover, farming on slopes tends to be associated with greater diversity in both crops and varieties (Gauchan and Smale, 2007). In these cases, tolerant varieties are planted where there would otherwise be no vegetative cover, and multiple varieties are planted to best match soil type. This provides for a more continuous planted biomass, and so avoids or decreases soil

erosion (and at the same time enhances the soil's ability to sequester carbon).

There are well-documented cases where the low fruit set of crops – and the resulting reduction in yield – has been clearly attributed to pollinator impoverishment. As most temperate and tropical fruit trees are obligatory outcrossers, and rely on insects or small animals for pollination, there is great potential for enhancing the role of the varietal diversity of the fruit trees themselves in promoting cross-hybridization and better fruit production. Studies have shown that strategic plantings, alternating different varieties in a chequerboard pattern for example, can optimize effective pollination visits to two varieties of different attractiveness and, at the same time, promote cross-hybridization and better fruit production (Kubišová and Háslbachová, 1991). In a similar approach, pollinator-attracting genotypes of certain crops have been explored as a management strategy for enhancing pollination services (Suso *et al.*, 2008), as genetic polymorphism in the reproductive characters of flowering plants can influence pollinator foraging (Cane and Schiffhauer, 2001). Diversity that promotes staggered flowering times among crop varieties has the potential to prolong season-long visitation by bees throughout the protracted flowering season (thus increasing the chances of pollinator population survival to the next growing season), as well as to increase the types of bees visiting at different times during the season, because several bee species are sensitive to climatic variation (Willmer *et al.*, 1994; Kremen *et al.*, 2002). In the Yucatan, Mexico, this management strategy is used with maize varieties; short-cycle maize and the more popular long-cycle maize are planted together in order to supply bees with pollen during the wet season and sustain the bee population until the next floral season (Tuxill, 2005).

Constraints and Policy Options

Though many of the management practices discussed here are both more environmentally sustainable and could result in beneficial economic returns, adoption is not guaranteed (see also Chapter 8). This can be due to limited access to information, to appropriate technologies or to finance (FAO, 2007). In addition, subsidies for agricultural production can lead to practices that degrade ecosystems. Other reasons for the non-adoption of sustainable technologies include inclusion in or exclusion from social networks (Warriner and Moul, 1992), land tenure (Tenge *et al.*, 2004) and sociocultural determinants.

Policy makers have an important role to play in safeguarding ecosystem services. Accounting for the benefits and costs of the full range of ecosystem services in policy making, and greater emphasis on natural resources and water use efficiency in food production, will promote better decision making that will lead towards more sustainable farming. Subsequently, coherence in cross-sector policies is fundamental to supporting collaboration among various stakeholders. Inter-sectoral collaboration at the ministerial level is essential for ensuring good ecosystem care, while providing the necessary food and services to communities. The need for coherence applies at the national level, between ministries of agriculture, the environment, water and natural resources; likewise, it applies in donor policy and, not least, between national governments and international institutions (Fresco, 2005).

Conclusions

To harness the full value of the ecosystem services that can be derived from sustainable water management practices linked to sustainable soil and biological diversity within agricultural ecosystems and their surrounding areas, a paradigm shift is needed in the way agriculture is carried out. This shift will require a move away from single solutions to production problems, towards risk reduction by creating insurance through a multitude of ways to better use soil, water and biotic resources that enhance ecosystem services. It will support the need for the enhanced capacity of natural resource managers to recognize, assist and create partnerships with small-scale farmers that adopt water, soil and biotic management methods – methods that will both reduce vulnerability in the production system and, at the same time, maintain productivity. The change will also require efforts to promote different norms among the consumers and retailers that support agricultural production systems, so that the vulnerability of these systems is reduced, together with continued productivity through enhanced ecosystem services. A change such as this will need to be supported by policies, legal measures and incentives that support production systems with less dependence on external inputs, and/ or wiser management of these resources.

References

Aguirre, J.A., Bellon, M.R. and Smale, M. (2000) A regional analysis of maize biological diversity in southeastern Guanajuato, Mexico. *Economic Botany* 54, 60–72.

Barrios, E., Sileshi, G.W., Shepherd, K. and Sinclair, F. (2012) Agroforestry and soil health: linking trees, soil biota and ecosystem services. In: Wall, D.H., Bardgett, R.D., Behan-Pelletier, V., Herrick, J.E., Jones, H., Ritz, K., Six, J., Strong, D.R. and van der Putten, W.H. (eds) *Soil Ecology and Ecosystem Services*. Oxford University Press, Oxford, UK, pp. 315–330.

Bennett, E.M., Peterson, G.D. and Levitt, E.A. (2005) Looking to the future of ecosystem services. *Ecosystems* 8, 125–132. doi:10.1007/s10021-004-0078-y

Bhogal, A., Nicholson, F.A. and Chambers, B.J. (2009) Organic carbon additions: effects on soil bio-physical and physico-chemical properties. *European Journal of Soil Science* 60, 276–286.

Bhogal, A., Nicholson, F.A., Young, I., Sturrock, C., Whitmore, A.P. and Chambers, B.J. (2011) Effects of recent and accumulated livestock manure carbon additions on soil fertility and quality. *European Journal of Soil Science* 62, 174–181.

Bindraban, P.S. *et al.* (2012) Assessing the impact of soil degradation on food production. *Current Opinion in Environmental Sustainability* 4, 478–488.

Boisvert, R.N. and Chang, H.H. (2006) *Multifunctional Agricultural Policy, Reduced Domestic Support, and Liberalized Trade: An Empirical Assessment for Taiwanese Rice.* Comprehensive Assessment of Water Management in Agriculture Research Report 14, International Water Management Institute (IWMI), Colombo. doi:10.3910/2009.379

Boody, G., Vondracek, B., Andow, D.A., Krinke, M., Westra, J., Zimmerman, J. and Welle, P. (2005) Multifunctional agriculture in the United States. *BioScience* 55, 27–38.

Bouwman, L., Goldewijk, K.K., Van Der Hoek, K.W., Beusen, A.H.W., Van Vuuren, D.P., Willems, J., Rufino, M.C. and Stehfest, E. (2011) Exploring global changes in nitrogen and phosphorus cycles in agriculture induced by livestock production over the 1900–2050 period. *Proceedings of the National Academy of Sciences of the United States of America.* doi 10.1073/pnas.1012878108; correction (2012) doi:10.1073/pnas.1206191109.

Brush, S. (1991) A farmer-based approach to conserving crop germplasm. *Economic Botany* 45, 153–165.

Cane, J. and Schiffhauer, D. (2001) Pollinator genetics and pollination: do honey bee colonies selected for pollen-hoarding field better pollinators of cranberry *Vaccinium macrocarpon. Ecological Entomology* 26, 117–123.

Deng, L., Shangguan, Z.-P. and Li, R. (2012) Effects of the grain-for-green program on soil erosion in China. *International Journal of Sediment Research* 27, 120–127. doi:10.1016/S1001-6279(12)60021-3

Dent, D. and Kauffman, J.H. (2007) Green water credits for rainwater management by farmers. *Global Water News* 4, 4–5. Available at: www.gwsp.org/fileadmin/downloads/GWSP_NL4_Internetversion.pdf (accessed December 2012).

Descheemaeker, K., Raes, D., Nyssen, J., Poesen, J., Haile M. and Deckers, J. (2009) Changes in water flows and water productivity upon vegetation regeneration on degraded hillslopes in northern Ethiopia: a water balance modelling exercise. *The Rangeland Journal* 31, 237–249. doi:10.1071/RJ09010

Descheemaeker, K., Amede, T. and Haileslassie, A. (2010a) Improving water productivity in mixed crop–livestock farming systems of sub-Saharan Africa. *Agricultural Water Management* 97, 579–586. doi:10.1016/j.agwat.2009.11.012

Descheemaeker, K., Mapedza, E., Amede, T. and Ayalneh, W. (2010b) Effects of integrated watershed management on livestock water productivity in water scarce areas in Ethiopia. *Physics and Chemistry of the Earth* 35, 723–729. doi:10.1016/j.pce.2010.06.006

Diaz, S. and Cabido, S. (2001) Vive la difference: plant functional diversity matters to ecosystem processes. *Trends in Ecology and Evolution* 16, 646–655.

Di Falco, S. and Chavas, J.P. (2007) On the role of crop biodiversity in the management of environmental risk. In: Kontoleon, A., Pascual, U. and Swanson, T.M. (eds) *Biodiversity Economics*. Cambridge University Press, Cambridge, UK, pp. 581–593.

Drechsel, P., Giordano, M. and Gyiele, L. (2004). *Valuing Nutrients in Soil and Water: Concepts and Techniques with Examples from IWMI Studies in the Developing World.* IWMI Research Report 82, International Water Management Institute, Colombo. doi:10.3910/2009.083

Dufour, A., Bartram, J., Bos, R. and Gannon, V. (eds) (2012) *Animal Waste, Water Quality and Human Health.* Published on behalf of the World Health Organization, Geneva, Switzerland, by IWA Publishing, London.

Dunn, H. (2011) *Payments for Ecosystem Services.* Paper 4, Defra Evidence and Analysis Series, Department for Environment, Food and Rural Affairs, London.

Ecoagriculture Partners (2012) Ecoagriculture Partners: Landscapes for People, Food and Nature. Washington, DC. Available at: http://www.ecoagriculture.org (accessed February 2013).

Enters, T. (1998) Methods for the economic assessment of the on- and off-site impact of soil erosion. IBSRAM Issues in Sustainable Land Management No. 2, International Board for Soil Research and Management, Bangkok.

FAO (1995) *Sustainability Issues in Agricultural and Rural Development Policies, Trainer's Manual, Volume 1.* Food and Agriculture Organization of the United Nations, Rome.

FAO (2007) *The State of Food and Agriculture: Paying Farmers for Environmental Services.* FAO Agriculture Series No. 38, Food and Agriculture Organization of the United Nations, Rome. Available at: www.fao.org/docrep/010/a1200e/a1200e00.htm (accessed June 2011).

FAO and PAR (2011) *Biodiversity for Food and Agriculture. Contributing to Food Security and Sustainability in a Changing World. Outcomes of an Expert Workshop Held by FAO and the Platform on [for] Agrobiodiversity Research from 14–16 April 2010 in Rome, Italy.* Food and Agriculture Organization of the United Nations, Rome and Platform for Agrobiodiversity Research, Rome, Italy. Available at: http://agrobiodiversityplatform.org/files/2011/04/PAR-FAO-book_lr.pdf (accessed December 2012).

Finckh, M.R. (2003) Ecological benefits of diversification. In: Mew, T.W., Brar, D.S., Peng, S., Dawe, D. and Hardy, B. (eds) *Rice Science: Innovations and Impact for Livelihood. Proceedings of the International Rice Research Conference, September 16–19, 2002, Beijing, China.* International Rice Research Institute, Los Baños, Philippines and Chinese Academy of Engineering and Chinese Academy of Agricultural Sciences, Beijing, China, pp. 549–564.

Finckh, M.R. and Wolfe, M.S. (2006) Diversification strategies. In: Cooke, B.M. Gareth Jones, D. and Kaye, B. (eds) *The Epidemiology of Plant Disease*, 2nd edn. Springer, Dordrecht, the Netherlands, pp. 269–308.

Finckh, M., Gacek, E., Goyeau, H., Lannou, C., Merz, U., Mundt, C., Munk, L., Nadziak, J., Newton, A., de Vallavieille-Pope, C. and Wolfe, M. (2000) Cereal variety and species mixtures in practice, with emphasis on disease resistance. *Agronomie* 20, 813–837.

Foley, J.A. *et al.* (2005) Global consequences of land use. *Science* 309, 570–574. doi:10.1126/science.1111772

Francis, C.A. and Porter, P. (2011) Ecology in sustainable agriculture practices and systems. *Critical Reviews in Plant Sciences* 30, 64–73. doi:10.1080/07352689.2011.554353

Fresco, L.O. (2005) Water, food and ecosystems in Africa. *Spotlight Magazine*, 3 February 2005. Food and Agriculture Organization of the United Nations, Rome. Available at: www.fao.org/ag/magazine/0502sp1.htm (accessed December 2012).

Gauchan, D. and Smale, M. (2007) Comparing the choices of farmers and breeders: the value of rice landraces in Nepal. In: Jarvis, D.I., Padoch, C. and Cooper, H.D. (eds) *Managing Biodiversity in Agricultural Ecosystems.* Published for Bioversity International, Rome by Columbia University Press, New York, pp. 407–425.

German, L., Mansoor, H., Alemu, G., Mazengia, W., Amede, T. and Stroud, A. (2007) Participatory integrated watershed management: Evolution of concepts and methods in an ecoregional program of the eastern African highlands. *Agricultural Systems* 94(2), 189–204. doi:10.1016/j.agsy.2006.08.008

Gomiero, T., Pimentel, D. and Paoletti, M.G. (2011) Environmental impact of different agricultural management practices: conventional vs. organic agriculture. *Critical Reviews in Plant Sciences* 30, 95–124. doi:10.1080/07352689.2011.554355

Hajjar, R., Jarvis, D.I. and Gemmill-Herren, B. (2008) The utility of crop genetic diversity in maintaining ecosystem services. *Agriculture, Ecosystems and Environment* 123, 261–270. doi:10.1016/j.agee.2007.08.003

ITAIPU (2011) Cultivando Água Boa [Growing Good Water]. Available at: www2.itaipu.gov.br/cultivandoaguaboa/ (accessed October 2012).

Jackson, B., Pagella, T., Sinclair, F., Orellana, B., Henshaw, A., Reynolds, B., Mcintyre, N., Wheater, H. and Eycott, A. (2013) Polyscape: A GIS mapping framework providing efficient and spatially explicit landscape-scale valuation of multiple ecosystem services. *Landscape and Urban Planning* 112, 74–88. doi:10.1016/j.landurbplan.2012.12.014

Jansen, H., Hengsdijk, H., Legesse, D., Ayenew, T., Hellegers, P. and Spliethoff, P. (2007) *Land and Water Resources Assessment in the Ethiopian Central Rift Valley. Project: Ecosystems for Water, Food and Economic Development.* Alterra Report 1587, Alterra Research Institute, Wageningen University, Wageningen, the Netherlands. Available at: http://content.alterra.wur.nl/Webdocs/PDFFiles/Alterrarapporten/AlterraRapport1587.pdf (accessed December 2012).

Jarvis, D.I., Brown, A.D.H., Imbruce, V., Ochoa, J., Sadiki, M., Karamura, E., Trutmann, P. and Finckh, M.R. (2007) Managing crop disease in traditional agroecosystems: the benefits and hazards of genetic diversity. In: Jarvis, D.I., Padoch, C. and Cooper, H.D. (eds) *Managing Biodiversity in Agricultural Ecosystems.* Published for Bioversity International, Rome by Columbia University Press, New York, pp. 292–319.

Kelsey Jack, B., Kousky, C. and Sims, K.R.E. (2008) Designing payments for ecosystem services: lessons from previous experience with incentive-based mechanisms. *Proceedings of the National Academy of Sciences of the United States of America* 105, 9465–-9470. doi:10.1073/pnas.0705503104

Kerr, J. (2002) Sharing the benefits of watershed management in Sukhomajri, India. In: Pagiola, S., Bishop, J. and Landell-Mills, N. (eds) *Selling Forest Environmental Services: Market-based Mechanisms for Conservation and Development.* Earthscan, London.

Kremen, C., Williams, N. and Thorp, R. (2002) Crop pollination from native bees at risk from agricultural intensification. *Proceedings of the National Academy of Sciences of the United States of America* 99, 16812–16816.

Kubišová, S. and Háslbachová, H. (1991) Pollination of male-sterile green pepper line (*Capsicum annuum* L.) by honeybees. In: van Heemert, K. and de Ruijter, A. (eds) *VI International Symposium on Pollination, Tilburg, the Netherlands, August 1990. Acta Horticulturae* 288, 364–370.

Liu, B.M., Abebe, Y., McHugh, O.V., Collick, A.S., Gebrekidan, B. and Steenhuis, T.S. (2008) Overcoming limited information through participatory watershed management: case study in Amhara, Ethiopia. *Physics and Chemistry of the Earth* 33, 13–21. doi:10.1016/j.pce.2007.04.017

Masse, D.I., Talbot, G. and Gilbert, Y. (2011) On farm biogas production: a method to reduce GHG emissions and develop more sustainable livestock operations. *Animal Feed Science and Technology* 166, 436–445.

Matsuno, Y., Ko, H.S., Tan, C.H., Barker, R. and Levine, G. (2002) *Accounting of Agricultural and Nonagricultural Impacts of Irrigation and Drainage Systems: A Study of Multifunctionality in Rice.* IWMI Working Paper 43, International Water Management Institute (IWMI), Colombo.

Mello I. and van Raij, B. (2006) No-till for sustainable agriculture in Brazil. *Proceedings of the World Association of Soil and Water Conservation* P1, 49–57.

Mirkena, T., Duguma, G., William, A., Wurzinger, M., Haile, A., Rischkowsky, B., Okeyo, A.M., Tibbo, M. and Solkner, J. (2011) Community-based alternative breeding plans for indigenous sheep breeds in four agroecological zones of Ethiopia. *Journal of Animal Breeding and Genetics* 2011, 1–10.

Molden, D. (ed.) (2007) *Water for Food, Water for Life: Comprehensive Assessment of Water Management in Agriculture.* Earthscan, London, in association with International Water Management Institute (IWMI), Colombo.

Muli, A. (2000) Factors affecting amount of water offered to dairy cattle in Kiambu District and their effects on productivity. BSc thesis, University of Nairobi (Range Management), Nairobi.

Mulumba, J.W., Nankya, R., Adokorach, J., Kiwuka, C., Fadda, C., De Santis, P. and Jarvis, P.I. (2012) A risk-minimizing argument for traditional crop varietal diversity use to recue pest and disease damage in agricultural ecosystems of Uganda. *Agriculture, Ecosystems and Environment* 157, 70–86. doi:10.1016/j.agee.2012.02.012

Nguyen-Khoa, S. and Smith, L.E.D. (2008) Fishing in the paddy fields of monsoon developing countries: re-focusing the current discourse on the 'multifunctionality of agriculture'. Keynote paper presented at the INWEPF-ICID Workshop of the RAMSAR COP10 Meeting, Changwon, Korea, 28 October–04 November 2008.

Oenema, O., Oudendag, D. and Velthof, G.L. (2007) Nutrient losses from manure management in the European Union. *Livestock Science* 112, 261–272. doi:10.1016/j.livsci.2007.09.007

Porter, J., Costanza, R., Sandhu, H., Sigsgaard, L. and Wratten, S. (2009) The value of producing food, energy, and ecosystem services within an agro-ecosystem. *Ambio* 38, 186–193.

Powell, M., Pearson, A.R. and Hiernaux, P.H. (2004) Crop-livestock interactions in the West African drylands. *Agronomy Journal* 96, 469–483. doi:10.2134/agronj2004.4690

Pradhan, N., Providoli, I., Regmi, B. and Kafle, G. (2010) Valuing water and its ecological services in rural landscapes: a case study from Nepal. *Mountain Forum Bulletin*, January 2010, 32–34. Available at: http://lib.icimod.org/record/14618/files/5475.pdf (accessed February 2013).

Puskur, R., Gebreselassie, S., Ayalneh, W., Yimegnuhal, A. and Peden, D. (2006) Integrating water harvesting and small scale dairy production: implications for livelihoods. Unpublished report, International Livestock Research Institute, Addis Ababa.

Rufino, M.C., Rowe, E.C., Delve, R.J. and Giller, K.E. (2006) Nitrogen cycling efficiencies through resource-poor African crop–livestock systems. *Agriculture, Ecosystems and Environment* 112, 261–282. doi:10.1016/j.agee.2005.08.028

Rufino, M.C., Tittonell, P., van Wijk, M.T., Castellanos-Navarrete, A., Delve, R.J., de Ridder, N. and Giller, K.E. (2007) Manure as a key resource within smallholder farming systems: analysing farm-scale nutrient cycling efficiencies with the NUANCES framework. *Livestock Science* 112, 273–287.

Sadiki, M. (2006) Diversity of Moroccan local faba bean landraces for reaction to drought stress. In: Jarvis, D., Mar, I. and Sears, L. (eds) *Enhancing the Use of Crop Genetic Diversity to Manage Abiotic Stress in Agricultural Production. Proceedings of a Workshop, 23–27 May 2005, Budapest, Hungary.* International Plant Genetic Resources Institute (IPGRI), Rome, pp. 11–17.

Sawadogo, M., Balma, D., Some, L., Fadda, C. and Jarvis, D. (2006) Management of the agrobiodiversity under the clinal variation of rainfall pattern in Burkina Faso: the example of okra drought resistance. In: Jarvis, D., Mar, I. and Sears, L. (eds) *Enhancing the Use of Crop Genetic Diversity to Manage Abiotic Stress in Agricultural Production. Proceedings of a Workshop, 23–27 May 2005, Budapest, Hungary.* International Plant Genetic Resources Institute (IPGRI), Rome, pp. 18–24.

Scherr, S.J. and McNeely, J.A. (2008) Biodiversity conservation and agricultural sustainability: towards a new paradigm of 'ecoagriculture' landscapes. *Philosophical Transactions of the Royal Society B* 363, 477–494. doi:10.1098/rstb.2007.2165

Schuman, G.E., Janzen, H.H. and Herrick, J.E. (2002) Soil carbon dynamics and potential carbon sequestration by rangelands. *Environmental Pollution* 116, 391–396. doi:10.1016/S0269-7491(01)00215-9

Sengupta, S., Mitra, K., Saigal, S., Gupta, R., Tiwari, S. and Peters, N. (2003) Developing markets for watershed protection services and improved livelihoods in India. Discussion Paper, Winrock International India, New Delhi and International Institute for Environment and Development, London (unpublished draft). Available at: http://pubs.iied.org/pubs/pdfs/G00399.pdf (accessed December 2012).

Shiferaw, B.A., Okello, J. and Reddy, R.V. (2009) Adoption and adaptation of natural resource management innovations in smallholder agriculture: reflections on key lessons and best practices. *Environment, Development and Sustainability* 11, 601–619.

Staal, S. *et al.* (2001) *Dairy Systems Characterisation of the Greater Nairobi Milk Shed.* Smallholder Dairy (R&D) Project Report, KARI/MoA/ILRI (Kenya Agricultural Research Institute/Ministry of Agriculture/International Livestock Research Institute) Collaborative Dairy Research Programme, ILRI, Nairobi.

Suso, M.J., Nadal, S., Roman, B. and Gilsanz, S. (2008) *Vicia faba* germplasm multiplication – floral traits associated with pollen-mediated gene flow under diverse between-plot isolation strategies. *Annuals of Applied Biology* 152, 201–208.

Swinton, S., Lupi, G., Robertson, P. and Landis, D. (2006) Ecosystem services from agriculture: looking beyond the usual suspects. *American Journal of Agricultural Economics* 88, 1160–1166.

Tarawali, S., Herrero, M., Descheemaeker, K., Grings, E. and Blümmel, M. (2011) Pathways for sustainable development of mixed crop livestock systems: taking a livestock and pro-poor approach. *Livestock Science* 139, 11–21. doi:10.1016/j.livsci.2011.03.003

Tenge, A.J., De Graaff, J. and Hella, J.P. (2004) Social and economic factors affecting the adoption of soil and water conservation in West Usambara highlands, Tanzania. *Land Degradation and Development* 15, 99–114. doi:10.1002/ldr.606

Thinlay, X., Finckh, M.R., Bordeosc, A.C. and Zeigler, R.S. (2000) Effects and possible causes of an unprecedented rice blast epidemic on the traditional farming system of Bhutan. *Agriculture, Ecosystems and Environment* 78, 237–248.

Thiombiano, L. and Meshack, M. (2009) *Scaling Up Conservation Agriculture in Africa: Strategy and Approaches.* Food and Agriculture Organization of the United Nations (FAO), Subregional Office for Eastern Africa, Addis Ababa. Available at: www.fao.org/ag/ca/doc/conservation.pdf (accessed December 2012).

Thurston, H.D., Salick, J., Smith, M.E., Trutmann, P., Pham J.L. and McDowell, R. (1999) Traditional management of agrobiodiversity. In: Wood, D. and Lenné, J.M. (eds) *Agrobiodiversity: Characterization, Utilization and Management.* CAB International, Wallingford, UK, pp. 211–243.

Tilman, D., Balzer, C., Hill, J. and Befort, B.L. (2011) Global food demand and the sustainable intensification of agriculture. *Proceedings of the National Academy of Sciences of the United States of America* 108, 20260–2026410. doi:1073/pnas.1116437108

TNC (The Nature Conservancy) (2008) Brazil: Plant a Billion Trees Campaign in the Atlantic forest. Available at: http://adopt.nature.org/plantabillion/brazil/ (accessed February 2013).

Trutmann, P., Fairhead, J. and Voss, J. (1993) Management of common bean diseases by farmers in the Central African highlands. *International Journal of Pest Management* 39, 334–342.

Tuxill, J. (2005) Agrarian change and crop diversity in Mayan milpas of Yucatan, Mexico: implications for on-farm conservation. PhD thesis, Yale University, New Haven, Connecticut.

UNDP, UNEP, World Bank and WRI (2000) *A Guide to World Resources 2000–2001: People and Ecosystems: The Fraying Web of Life.* United Nations Development Programme, United Nations Environment Programme, World Bank and World Resources Institute, Washington, DC.

UNFPA (2009) *The State of World Population 2009. Facing a Changing World: Women, Population and Climate.* United Nations Population Fund, New York. Available at: www.unfpa.org/swp/2009/ (accessed December 2012).

Vereijken, P. (2003) Transition to multifunctional land use and agriculture. *Netherlands Journal of Agricultural Science* 50, 171–179.

Waltner-Toews, D. (2009) Eco-health: a primer for veterinarians. *Canadian Veterinary Journal* 50, 519–521.

Warriner, G.K. and Moul, T.M. (1992) Kinship and personal communication network influences on the adoption of agriculture conservation technology. *Journal of Rural Studies* 8, 279–291. doi:10.1016/0743-0167(92)90005-Q

Watson, R.T., Noble, I.R., Bolin, B., Ravindranath, N.H., Verardo, D.J. and Dokken, D.J. (eds) (2000) *Land Use, Land Use Change and Forestry. A Special Report of the Intergovernmental Panel on Climate Change.* Cambridge University Press, Cambridge, UK.

Weltzien, E., Rattunde, H.F.W., Clerget, B., Siart, S., Toure, A. and Sagnard, F. (2006) Sorghum diversity and adaptation to drought in West Africa. In: Jarvis, D., Mar, I. and Sears, L. (eds) *Enhancing the Use of Crop Genetic Diversity to Manage Abiotic Stress in Agricultural Production. Proceedings of a Workshop, 23–27 May 2005, Budapest, Hungary.* International Plant Genetic Resources Institute (IPGRI), Rome, pp. 31–38.

Wilby, A. and Thomas, M. (2007) Diversity and pest management in agroecosystems – some perspectives from ecology. In: Jarvis, D.I., Padoch, C. and Cooper, H.D. (eds) *Managing Biodiversity in Agricultural Ecosystems.* Published for Bioversity International, Rome by Columbia University Press, New York, pp. 269–291.

Willmer, P., Bataw, A. and Hughes, J. (1994) The superiority of bumblebees to honeybees as pollinators: insect visits to raspberry flowers. *Ecological Entomology* 19, 271–284. doi:10.1111/j.1365-2311.1994.tb00419.x

Wolfe, M. (1985) The current status and prospects of multiline cultivars and variety mixtures for disease resistance. *Annual Review of Phytopathology* 23, 251–273.

Woomer, P.L., Touré, A. and Sall, M. (2004) Carbon stocks in Senegal's Sahel transition zone. *Journal of Arid Environments* 59, 499–510. doi:10.1016/j.jaridenv.2004.03.027

World Bank (2009) *Minding the Stock: Bringing Public Policy to Bear on Livestock Sector Development.* Report No. 44110-GLB, World Bank, Agriculture and Rural Development Department, Washington, DC.

WWF (2006) *Payments for Environmental Services. An Equitable Approach for Reducing Poverty and Conserving Nature.* World Wide Fund For Nature, Gland, Switzerland. Available at: http://www.unpei.org/PDF/ecosystems-economicanalysis/Payments-for-Ecosystem-Services.pdf (accessed February 2013).

Xu, Z., Bennett, M.T., Tao, R. and Xu, J. (2004) China's Sloping Land Conversion Programme four years on: current situation, pending issues. *International Forestry Review* 6, 317–326.

10 Water Management for Ecosystem Health and Food Production

Gareth J. Lloyd,[1]* Louise Korsgaard,[1]† Rebecca E. Tharme,[2] Eline Boelee,[3] Floriane Clement,[4] Jennie Barron[5] and Nishadi Eriyagama[6]

[1]*UNEP–DHI Centre for Water and Environment, Hørsholm, Denmark;* [2]*The Nature Conservancy (TNC), Buxton, UK;* [3]*Water Health, Hollandsche Rading, the Netherlands;* [4]*International Water Management Institute (IWMI), Kathmandu, Nepal;* [5]*Stockholm Environment Institute, University of York, UK and Stockholm Resilience Centre, Stockholm University, Stockholm, Sweden;* [6]*International Water Management Institute (IWMI), Colombo, Sri Lanka*

Abstract

The integrated, efficient, equitable and sustainable management of water resources is of vital importance for securing ecosystem health and services to people, not least of which is food production. The challenges related to increasing water scarcity and ecosystem degradation, and the added complexities of climate change, highlight the need for countries to carefully manage their surface water and groundwater resources. Built upon the principles of economic efficiency, equity and environmental sustainability, integrated water resources management (IWRM) can be shaped by local needs to maximize allocative efficiency and better manage water for people, food, nature and industry. However, the flexibility of the approach means that it is interpreted and applied in ways that prioritize and address immediate challenges created by demographic, economic and social drivers, often at the expense of environmental sustainability – and hence also of long-term food security. The need to more explicitly include ecosystems in water management practices and safeguard long-term food security can be addressed partly by refining the notion of 'water for food' in IWRM as 'water for agroecosystems'. This would also serve to eliminate much of the current dichotomy between 'water for food' and 'water for nature', and deliver a more balanced approach to ecosystem services that explicitly considers the value and benefits to people of a healthy resource base. The adoption of an ecosystem services approach to IWRM, and incorporation of environmental flows as a key element, can contribute to long-term food security and ecosystem health by ensuring more efficient and effective management of water for agroecosystems, natural systems and all its other uses.

* E-mail: gjl@dhigroup.com
† E-mail: lok@dhigroup.com

Background

The water cycle enables ecosystems to provision goods such as food, fuel and timber; to regulate and support the environment and its biological diversity; and to provide for cultural services and fundamental ecological processes (Millennium Ecosystem Assessment, 2005; Gordon *et al.*, 2010; Chapter 3). Thus, ecosystem integrity and long-term health are at the very centre of sustainable food production, and efficient, equitable and sustainable management of water resources is crucial for both ecosystem health and food production. The challenges related to increasing water scarcity and climate change (Chapters 2 and 5), highlight the need to achieve the greatest possible water use efficiency in an economically, politically, environmentally and socially acceptable manner. Several options for improving the efficiency of water use for both food production and the maintenance of ecosystem services have already been discussed, and the concept of environmental flows has been introduced (respectively, in Chapters 5, 8 and 9). Arguably, the more challenging issue has been how to implement these advances (e.g. Naiman *et al.*, 2002; Rowlston and Tharme, 2008; Le Quesne *et al.*, 2010) and to enhance water-use efficiency, while increasing food production and simultaneously meeting ecosystem needs. In many instances, the need to address this issue stems from the fact that water savings from agricultural efficiency are channelled back into further agricultural production, rather than to securing adequate long-term ecosystem health.

Historically, attempts to balance water for food, people, nature and industry have typically led to the further entrenchment of silo-like, sectoral policy making and planning at national government level, the result of which is fragmented water governance that takes little or no account of water uses beyond the interests and jurisdiction of individual sectors. Recognition of the lack of sustainability of such an approach under conditions of water stress, competing demands and high variability in water availability has resulted in an explosion of interest in integrated water resources management (IWRM) in recent years (e.g. Snellen and Schrevel, 2004). Since the adoption of the Agenda 21 principles in 1992, an increasing number of nations have introduced national policies that adhere to the principles of IWRM and include associated strategies (UN Water, 2012). In a global survey with 133 country responses, more than 70% stated that water management had been introduced in national policy and legislation to actively account for water resources development, impacts by other sectors and multiple demands (UN Water, 2012). Similar evidence exists for countries in sub-Saharan Africa over the last 10 years. In a survey of 24 eastern and southern African countries, it was clear that most countries had put into place the enabling conditions in terms of policies founded on the principles of IWRM (GWP Eastern Africa and GWP Southern Africa, 2010). The operationalization of IWRM still lags behind though owing to resources gaps in finance, and in human and institutional capacity.

Refining Integrated Water Resources Management (IWRM)

IWRM can be described as 'the coordinated development and management of water, land and related resources, in order to maximize the resultant economic and social welfare in an equitable manner without compromising the sustainability of vital ecosystems' (GWP Technical Advisory Committee, 2000). Built upon principles of economic efficiency, social equity and environmental sustainability, sometimes referred to as the 'three Es', the IWRM approach offers the possibility of taking into account multiple economic, social and environmental needs. It takes the form and function of an all-encompassing management framework that can be used to consider and apply regulatory instruments, and to assimilate other practical measures that address water resources management. A good introduction to IWRM for policy makers and practitioners is the GWP (Global Water Partnership) ToolBox (GWP Toolbox, 2008).

Key to IWRM is an inter-sectoral approach that strives to ensure effective coordination of all sectors and uses of water; this is the 'IWRM comb' that is shown in Fig. 10.1. For example,

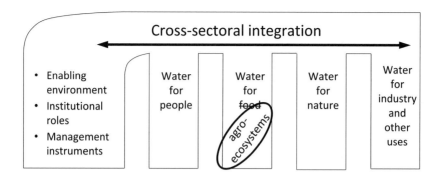

Fig. 10.1. The integrated water resources management (IWRM) comb (after GWP Technical Advisory Committee, 2000). Note: in this book, it is proposed to refine 'water for food' to 'water for agroecosystems', as discussed in the section entitled 'An Ecosystem Services Approach to Water Management' and shown in this figure.

planners for domestic water supply and sanitation (water for people), for irrigation and fisheries (water for food), for nature conservation (water for nature) and so on, must take other users' needs into consideration, particularly in terms of water allocation and the resulting impacts of allocation decisions. Management coordination based on a hydrological unit such as a lake, river or aquifer, rather than on political boundaries that may divide bodies of water, is another central aspect of IWRM. The combination of inter-sectoral and basin approaches makes IWRM suitable for efficient management of water in landscapes of various natural and agricultural ecosystems.

Some practitioners and scientists have criticized IWRM as being, for example, difficult to implement, insensitive to cultural differences and as not sufficiently encompassing emerging issues, such as climate change and water security (e.g. Biswas, 2004; Rahaman and Varis, 2005; Matz, 2008; Medema *et al.*, 2008; Chéné, 2009; Saravanan *et al.*, 2009).

However, in reality, IWRM plans and practices applied at regional, national and local levels are heavily influenced by local circumstances, requirements and interpret-ations (UN Water, 2012). For example, some stakeholders may refuse or be refused the opportunity to engage in an integrated management approach, and some plans may be developed based upon administrative borders, such as a city, or for a specific purpose, such as a flood situation, rather than on a specific hydrological unit, but the result may still be regarded by those involved as IWRM. While few would disagree that the operational reality of IWRM is highly complicated in trans-boundary situations (where various countries, states or regions have their own agendas and may be reluctant to cooperate with each other), many practitioners would agree that IWRM is just a tool, and it is the responsibility of those involved to determine how it should be used. Because of its flexibility and inclusiveness, IWRM is seen as a key prerequisite for ensuring climate resilience and water security (e.g. Kundzewicz *et al.*, 2007; WRG, 2009; AMCOW, 2012).

International efforts are currently being made to try to address the apparent conflict between short-term economic growth and sustainable water resource management by growing calls for green growth and green economy strategies that build upon the foundations of sustainable development (UNEP, 2011). In terms of future scenarios, the challenge that an increasing number of both developed and developing countries will face is how to reconcile a growing gap between water demands and available supplies in a way that meets their development objectives in a cost-effective way (WRG, 2009).

It is not unusual for political decision makers to work with operational planning horizons based on periods of no more than 5 years – what may or may not happen in 100, 50 or even 20 years is beyond their direct control.

As a result, priorities are typically shaped by the immediate challenges created by demographic, economic and social drivers; these, in turn, colour decisions regarding allocation efficiencies, and concerns about the environment are subordinated. From a resource policy and planning perspective, it is hence important to recognize the broad objectives that lie behind the promotion of sustainable or efficient water management through the adoption and application use of IWRM. For example, at the national level, countries invariably have numerous economic, social, environmental and political demands and counter-demands for multiple goods and services that require water as an input. Dealing with trade-offs and finding synergies between water for food and for other ecosystem services, as well as maintaining ecosystem integrity, is a huge challenge.

One way to address this challenge is through the application of a range of supply-side measures, such as: the development and operation of reservoirs and dams; improved maintenance of systems (including leakage control); rainwater harvesting; reuse/recycling of water; the development of surface and groundwater resources; and the application of water transfers. These measures increase the available resources; for efficiency, demand-side measures need to be applied as well.

Increasing Use Efficiency Through Demand Management and Allocative Efficiency

Considered in its most basic form, the term 'water use efficiency' assesses the amount of water needed to produce a given unit of any good or service (e.g. Seckler *et al.*, 2003). As discussed in Chapter 8, water use efficiency usually takes into account the water input, whereas water productivity uses the water consumption in its calculation, although the terms are often used interchangeably.

Minimizing the amount of water needed (reducing the demand) for the same outputs will result in greater efficiency. The aim is not always to reduce water use, but rather to optimize its utilization. From a food production point of view, much of the attention in the area of water use efficiency is given to how to maximize the amount of material produced per unit of water (thereby increasing 'water productivity', as discussed in Chapter 8). Sharma *et al.* (2010) combine analyses of water productivity, poverty linkages and institutional constraints to generate a series of recommendations for better integrated water management in the Indus and Ganges Plains of India. From the standpoint of ecosystem health and services provision, the aim of water use efficiency is to optimize the provision of a range of ecosystem services for a given amount of water and to maintain ecological integrity (e.g. through environmental flow provision). As with food production, it is crucial for such optimization that water is provided at the right time and in the right amount and quality.

For certain water uses, such as agriculture, industry and cities, water demand management is an effective means of increasing water use efficiency. The ultimate benefits of water demand management can be expressed in different ways: as gains yielded by increased economic efficiency of water use; as avoided losses resulting from current or future droughts, or from environmental degradation or ecosystem sustainability; and as avoided or postponed capital costs of enhanced water production. These benefits are complementary, but may not necessarily reinforce one other. Where current water supply meets the demand under normal conditions, the water demand management policies can create 'buffer' capacity against periods of below-normal water availability and thus help to avoid some of the costs inflicted by drought. Finally, where some water demands cannot be satisfied, such as in drylands (Chapter 6), water demand management can help to achieve the production of more value from the available water.

Representative demand-side measures that can contribute to water efficiency include:

- The application of economic and market-based instruments to motivate desired decision making, such as water tariff schemes with increasing rates based on volume used.
- The introduction of technologies and methods to increase water utility, such as the use of treated municipal wastewater for irrigation.

- The application of regulatory instruments that can be used to guide users, such as laws on the quantity and timing of abstractions.
- Awareness raising and capacity building instruments, such as information campaigns that inform users about the consequences of their actions or inactions.

Where matching demand with supply is not possible, allocative efficiency, a form of demand management, may be adopted. The goal of allocative efficiency is to maximize consumer satisfaction from available resources (Economic Glossary, 2012). IWRM is a useful tool for facilitating allocative efficiency, as its application provides the means by which various uses can be weighed and compared. In theory, who gets which water in what quantities and when is regulated by principles relating to economic efficiency, social equity and environmental sustainability – the 'three Es'. However, as noted above, in real life IWRM is interpreted and applied in multifarious ways, so its application is not always in harmony with the 'three Es'. This creates another set of challenges and raises questions on what is included and what gets left out – and on what basis such decisions are made.

While demand management measures applied through IWRM may be useful for increasing water use efficiency for economic sectors in the short term, beyond the textbooks these measures are not yet adequately addressing the vital role of ecosystems in sustainable water management and food production. There is a need to more explicitly include ecosystems in demand management practices.

An Ecosystem Services Approach to Water Management

Regardless of the overall framework for water resources management, be it IWRM or some other, there is growing recognition that more practical approaches to the fundamental issue of ecosystem management must be employed to support food production, ecosystem resilience and environmental sustainability (Molden, 2007). Healthy ecosystems provide a

wide range of valuable services (Millennium Ecosystem Assessment, 2005), and better ecosystem management can benefit agriculture and improve system water productivity in several ways (Chapter 3). Increased yields in resource-conserving agriculture can go hand in hand with reduced environmental impacts through increased water use efficiency and productivity, improved water quality and increased carbon sequestration. Balancing the goals of agricultural ecosystems with landscape ecosystem services can produce synergies and improve overall water productivity (Keys *et al.,* 2012). Water management that mimics natural water storage can improve agroecosystem water use at the same time as maintaining hydrological links with the surrounding landscape; this, in turn, preserves the water needed for additional ecosystem services (Keys *et al.*, 2012).

An integrated approach to land, water and ecosystem management could be based on IWRM (Falkenmark, 2003), could incorporate elements of the ecosystem services framework (ESF), and could benefit from a multiple-use water services (MUS) approach (van Koppen *et al.*, 2006, 2009). The three approaches are integrative by nature, and promote a more comprehensive view and analysis of water resources and uses, although they tend to be applied at different scales and with different entry points. For example, MUS is applied at the local level and with a focus on water supply infrastructure, IWRM starts with higher level policies, institutions or organizations, and ESF addresses ecosystems at the basin scale (Nguyen-Khoa and Smith, 2010).

More specific policy options and management approaches can help to strike a balance between increased food production and the preservation of ecosystems (Gordon *et al.*, 2010). For example, improved management practices on agricultural lands can increase the efficiency with which water is used to produce food, thereby allowing the opportunity for securing environmental flows with the saved water. Shifting from monocropping to multifunctional agroecosystems can create synergies among ecosystem services, meaning that all of the services are valued and cared for rather than just the crop yield output and its associated water productivity (Fig. 10.2)

(Molden *et al.*, 2007; Nguyen-Khoa and Smith, 2008).

The conversion and integration of an agricultural production system into a multifunctional agroecological landscape that delivers more balanced combinations of ecosystem services will take time, even if such a conversion is immediately biophysically practicable and socially acceptable. It involves not only the management of water and other natural resources such as land, but also an integrated approach at landscape or basin level (this is discussed in more detail in Chapter 11). In the interim, the value of ecosystem services delivered by changes in agricultural practices can increase substantially through measures to increase water and land productivity, and interventions that support specific ecosystem functions (Molden, 2007).

So far, the focus of water management, including IWRM, has mostly been on planning, allocating and managing surface water resources for irrigation (agriculture), energy (hydropower), industry and domestic water supply, while recognizing the need to safeguard environmental flows for aquatic ecosystem functions in rivers, lakes, estuaries and other wetlands. However, water for irrigation is better dealt with as water for agroecosystems (Fig. 10.1), and water for nature (environmental flows) should be valued and managed on equal terms with other water uses. Furthermore, key ecosystem services depend on water in the soil profile and the aquifers that support terrestrial ecosystems. As a consequence of this, water resource management needs to adopt an ecosystem services approach, and to incorporate environmental flows and include soil water alongside surface water needs. Reconsidering the 'water for food' tooth in the IWRM comb (Fig. 10.1), and applying it as 'water for agroecosystems', would be a way to

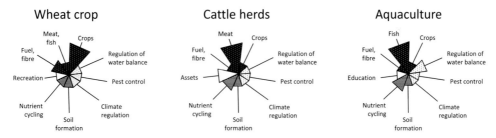

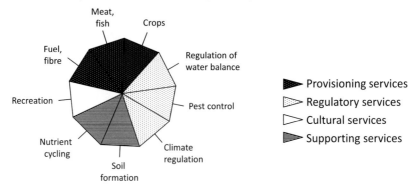

Fig. 10.2. Managing water for multifunctional agroecosystems would help a more balanced provision of provisioning, regulatory, cultural and supporting ecosystem services than single cropping (monocropping), extensive herding or peri-urban aquaculture (umbrella shape adapted from Molden, 2007; and Gordon *et al.*, 2010).

eliminate much of the current, somewhat divisive dichotomy between 'water for food' and 'water for nature' (Fig. 10.3). Thus, it would help to deliver a more balanced approach to ecosystem services that explicitly considers the value and benefits to people of a healthy resource base.

A major challenge to adopting an ecosystem services approach to water management is that the role and valuation of water in regulatory and supporting services remains poorly understood (Chapter 4), both in agroecosystems and in non-agricultural ecosystems, particularly with respect to soil- and groundwater-dependent systems. Moreover, water and accessible biomass together comprise an estimated 99% of all provisioning services (Weber, 2011). So even if there is a deliberate and increased emphasis on applying a policy of truly integrated management, this may not be sufficient to ensure that all or most of the desired ecosystem services are accounted for. It is, therefore, important to encourage the use

of adaptive management and adopt the precautionary principle when planning sustainable water management practices. Adaptive management, taking into account the adaptive capacity of the water resources themselves (precipitation, surface water and groundwater), as well as the adaptive capacity of their governing institutions (Pahl-Wostl *et al.*, 2007; Pahl-Wostl, 2009), is also key to responding to the implications of climate and other environmental changes for water resources and ecosystems.

An important implication of adopting an ecosystem approach is that, in agroecosystems, more so than in natural ecosystems, water requirements will change according to societal decisions on the extent to which water use is to be optimized for the full range of ecosystem services or, more typically, and often at greater risk to ecosystem integrity, maximized for particular combinations of services. In the same way, society ultimately decides the future level of health at which any natural ecosystem

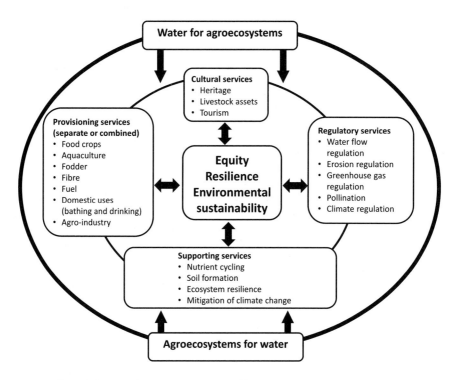

Fig. 10.3. Water for multifunctional agroecosystems would bring more equity, environmental sustainability and economic efficiency.

should be managed (Poff *et al.*, 2010). It is then a scientific question as to how much water is needed to achieve that particular level of health, and what the implications might be of not fully satisfying ecosystem water needs. Environmental flow assessments are an essential part of answering this question.

Applying Environmental Flows: Securing Water for Ecosystems

Water resources managers and scientists are increasingly integrating the concept and practices of environmental flows (Chapter 5) into IWRM, thereby increasing its likely uptake by other national, state and international actors. Such uptake is more likely to succeed where regulatory, economic and other market-based instruments, as well as awareness and capacity building, are applied within the IWRM framework to encourage greater water efficiency by planners and users.

Environmental flows may be thought of within an IWRM context in terms of 'environmental demand', similar to the way in which agricultural, industrial or domestic water demand are considered (Smakhtin and Eriyagama, 2008). These flows are aimed at maintaining an ecosystem in, or restoring it to, some scientifically defensible, societally prescribed or negotiated condition, also referred to as a 'desired future state', an 'environmental management class', an 'ecological management category' or a 'level of environmental protection' (e.g. DWAF, 1997; Acreman and Dunbar, 2004). In this way, environmental flows are commonly envisaged and approached as a negotiated trade-off, compromise or balanced optimization between objectives for river basin development on the one hand, and the maintenance of natural ecosystem integrity and biodiversity on the other (Naiman *et al.*, 2002; Postel and Richter, 2003).

The Global Environmental Flows Network has focused even more strongly on the connection with 'water for ecosystem services', defining environmental flows as 'the quantity, quality and timing of water flows required to sustain ecosystem services, in particular those related to downstream wetlands and aquatic

ecosystems and the human livelihoods and well-being that depend on these ecosystems' (adapted from eFlowNet, 2010). In that sense, agroecosystems could also be integrated into the ecosystems served by environmental flows. Korsgaard *et al.* (2008) developed a Service Provision Index (SPI) that links ecosystem services to flows, and allows for the valuation of environmental flows in socio-economic terms; this could potentially be used to more effectively integrate environmental flows into IWRM. Thus, values are put on ecosystem services served by environmental flows in the same way as they are put on ecosystem services (beyond food production) from agro-ecosystems. The increasing application of environmental flow assessments is making the vital connection between ecosystems and environmental flows explicit (Tharme, 2003).

The importance of the entire range of daily, seasonal and inter-annual variations in water flows (or levels) in sustaining the complete native biodiversity and integrity of aquatic ecosystems is well established (Poff *et al.*, 1997; Bunn and Arthington, 2002). Maintaining this full spectrum of naturally occurring flows and their inherent pattern of variability in a river (or other water body) is, however, often not feasible given the various competing sectoral demands associated with water resources development (for domestic supply, irrigation, flood control, hydropower, navigation, etc.), as well as changes in catchment land use and climate. With increasing alteration of the water flow regime from its natural pattern comes increasing ecological risk (Richter, 2009; Poff *et al.*, 2010). Hence, the higher the level and degree of assurance of ecosystem health and delivery of ecosystem services that are required, the more water will need to be reserved or allocated – as part of water resources planning – for maintaining ecosystem condition, and the more the system's flow magnitude, timing and pattern of variability will need to be preserved.

Many methods for environmental flow assessment that directly or indirectly encompass the above tenets have been developed over the years (e.g. Tharme, 2003; Acreman and Dunbar, 2004; IWMI, 2007). They differ significantly in their required information and other resource needs and, therefore, in the commensurate degree of resolution and confidence in their recommendations, and level of water resource planning or management for which they are most suited. Moreover, the majority of approaches to date have been applied for individual rivers, reaches or infrastructure projects, rather than for river systems or multiple projects at the whole-basin scale (Poff *et al.*, 2010).

Rapid planning (desktop) methods, typically of the lowest resolution and confidence, are based primarily on hydrological indices derived from the analysis and characterization of hydrological time series (e.g. Tennant, 1976; Hughes and Hannart, 2003; Smakhtin and Anputhas, 2006); in recent years, increasing effort has been dedicated to using more ecologically relevant flow indices (Tharme, 2003). Other approaches, such as higher confidence holistic methods, follow a rigorous protocol that typically addresses diverse ecohydrological and social components and processes, involves significant fieldwork and time, and employs a multidisciplinary panel of experts to derive the environmental flows needed for the ecosystem and for any directly dependent communities (e.g. Arthington, 1998; King *et al.*, 2003; Esselman and Opperman, 2010; see also Box 10.1). These approaches also rely on monitoring and adaptive management of the implemented flow recommendations in order to ensure that water management objectives are met for all water users (Konrad *et al.*, 2011).

Until recently, few countries, states or basin agencies had initiated environmental flow determinations at the river network or basin level, or at even broader scales, arguably because the groundwork necessary for such an approach was not yet laid. With the emergence of the ELOHA (ecological limits of hydrologic alteration) framework for assessing environmental flow needs in a large basin or region, particularly when in-depth studies cannot be performed for all its rivers (Arthington *et al.*, 2006), it is now possible to set environmental flow standards rapidly across large geographies (see Poff *et al.*, 2010). At present applied largely within the USA (see Box 10.2) and Australia, ELOHA is fast gaining traction in other places, such as Latin America, where the need for greater environmental

sustainability in basin water management is outpacing project-specific flow assessments.

Regional scale environmental flow assessments at whole basin, state or even country scales, often seem to promote more rapid and deeper engagement with national policy and regulatory frameworks and basin water resource management processes (as in the example in Box 10.2) than those at single project or site level. To date though, two of the major bottlenecks for the successful implementation of environmental flows, regardless of the scale at which environmental flows are determined, remain the inadequate involvement of stakeholders throughout the process and the lack of appropriate governance structures for effective implementation (Poff *et al.*, 2003; Le Quesne *et al.*, 2010). Recognition of this deficiency in water governance (Pahl-Wostl, 2009), coupled with inadequate inclusion to date of environmental flows into those global water assessments commonly used to examine future scenarios for water and food security, has resulted in various projects and programmes advocating further integration of these elements, so that true sustainability can be achieved in IWRM. An example is given by the Global Water System Project (GWSP; see Alcamo *et al.*, 2005) and its Global Water Needs Initiative (GWNI; see GWSP, 2013). Such initiatives continue to build on earlier work to address environmental water scarcity at a global scale – work which illustrated that even with the inclusion of environmental flow estimates of the order of only 20–50% of the mean annual flow in a river basin, large parts of the world already are, or will soon be, environmentally water stressed (e.g. Smakhtin *et al.*, 2004), so placing long-term resource sustainability at risk. However, this might not be the case if supporting and regulating ecosystem services in agroecosystems are enhanced through IWRM.

Conclusions

Built as it is upon the principles of economic efficiency, equity and environmental sustainability, integrated water resources management (IWRM) offers a comprehensive and adaptive management framework to support water

Box 10.1. Adopting a scenario-based approach to environmental flows in Tanzania. An example based on the Pangani River Basin Management Project (PRBMP) (IUCN, 2010; King *et al.*, 2010; PRBMP, 2010).

The Pangani River Basin covers about 43,650 km^2, mostly in Tanzania, with approximately 5% in Kenya. Flows in the basin have been reduced from several hundred to less than 40 m^3/s, as a result of largely uncontrolled irrigation and urban water demand. The remaining water is seriously over-allocated, with shortages affecting all water users – from mid-basin irrigators, to electricity producers further downstream, to coastal fisher communities with declining fish stocks owing to saline intrusion; conflicts are thus on the rise among the various sectors.

The International Union for the Conservation of Nature (IUCN), through its Water and Nature Initiative (WANI; see Smith and Cartin, 2011), started the multi-partner Pangani River Basin Management Project (PRBMP) in 2001 in order to improve management of the basin's water resources and to reduce the conflicts that were arising. The project aimed to: (i) assess environmental flow requirements to effectively conserve the basin's natural resources; (ii) establish fora for community participation in water management; and (iii) raise awareness about climate change impacts and adaptation strategies.

The project's flow assessment, undertaken in 2004–2008, used a modified Downstream Response to Imposed Flow Transformations process (DRIFT; see King *et al.*, 2003), and involved field and desktop work by a multidisciplinary expert group. Fifteen development scenarios and their associated flow scenarios were evaluated, and three reports were generated: 'state-of-the-basin'; 'flow assessment-scenario evaluation decision support system (DSS)'; and 'water allocation scenarios'. The results are currently being presented to stakeholders at all levels, with particular emphasis on the Pangani Basin Water Board (formerly Office), the governmental organization responsible for allocating water in the basin. Consultations with stakeholders are intended to raise awareness of the water issues in the basin, help select the best development path for the river and facilitate the integration of the selected environmental flow scenario into an integrated water resources management (IWRM) plan for the basin.

Box 10.2. Basin to statewide application of ELOHA in Colorado, USA: the Watershed Flow Evaluation Tool (Sanderson *et al.*, 2011).

To meet the need for regional flow management that addresses environmental sustainability in Colorado State, USA, the ELOHA (Ecological Limits Of Hydrologic Alteration) framework (Poff *et al.*, 2010) was applied to develop a Watershed Flow Evaluation Tool (WFET) for estimating flow-related ecological risk at a regional scale. The WFET entails: (i) modelling natural and developed daily streamflows; (ii) analysing the resulting flow time series; (iii) describing the relationships between river attributes and flow metrics (flow–ecology relationships); and (iv) mapping of flow-related risk for key in-stream and riparian biota. Two watersheds with differing geomorphic settings and data availability were studied, and the WFET was successfully implemented to assess basin flow-related ecological risk in one of them; active channel change and limited data precluded a successful application in the second basin. The WFET will be further used in Colorado to evaluate the risk of impacts on river ecosystems under future climate change, and to evaluate and balance ecosystem needs at the large scale within water development scenarios, such as for municipal water supply or energy development.

management for healthy ecosystems and food security. Associated economic and market-based, regulatory, awareness and capacity building instruments can be applied to manage demand and encourage greater water efficiency by planners and users.

As the focus of IWRM so far has mostly been on planning, allocating and managing surface water resources for irrigation, industry and water supply, there are good opportunities to recognize and embrace the need to safeguard environmental water for aquatic ecosystem health and long-term resiliency. The provision of key ecosystem services depends on adequate surface water, water in the soil profile and the aquifers of groundwater-dependent wetland and terrestrial ecosystems. Consequently, water resource management needs to adopt an ecosystem services approach that incorporates all elements of the water resource and give due attention to the value of allocating water for ecosystems – agroecosystems and non-agricultural or natural ecosystems alike. Better ecosystem management can, in turn, benefit food production and improve system water productivity in several ways.

To reflect this more directed focus on ecosystems, it is proposed to rephrase the 'water for food' tooth in the IWRM comb to 'water for agroecosystems'. This approach will avoid much of the current dichotomy between 'water for food' and 'water for nature' (or environmental flows) and help to deliver more balanced suites of ecosystem services, including those essential for food security.

The concept of environmental flows provides a basis for calculating the amount of water (quantity, quality and timing) required to sustain ecosystems and safeguard their services to people. This can also be applied to the 'water for agroecosystems' tooth in the IWRM comb. Water resource managers are increasingly applying the concept of environmental flows to IWRM and adopting the associated best practices, thereby increasing its likely uptake by other national and international actors.

To conclude, managing water efficiently for agroecosystems, nature and all other water uses by incorporating environmental flows and adopting an ecosystem services approach to IWRM can contribute to basin water sustainability, long-term food security and ecosystem health.

References

Acreman, M. and Dunbar, M.J. (2004) Defining environmental river flow requirements – a review. *Hydrology and Earth System Sciences* 8, 861–876. Available at: www.hydrol-earth-syst-sci.net/8/861/2004/hess-8-861-2004.pdf (accessed May 2011).

Alcamo, J., Grassl, H., Hoff, H., Kabat, P., Lansigan, F., Lawford, R., Lettenmaier, D., Lévêque, C., Meybeck, M., Naiman, R., Pahl-Wostl, C. and Vörösmarty, C. (2005) *The Global Water System Project: Science Framework and Implementation Activities.* ESSP Report No. 3, Earth System Science Partnership (International Geosphere–Biosphere Programme, International Human Dimensions Programme on Global Change, World Climate Research Programme and DIVERSITAS) and GWSP Report No.1, Global Water System Project, GWSP International Project Office, Bonn, Germany.

AMCOW (2012) *Status Report on the Application of Integrated Approaches to Water Resources Management in Africa.* African Ministers' Council on Water, Abuja, Nigeria.

Arthington, A.H. (1998) *Comparative Evaluation of Environmental Flow Assessment Techniques: Review of Holistic Methodologies.* Occasional Paper 26/98, Land and Water Resources Research and Development Corporation, Canberra, Australia.

Arthington, A.H., Bunn, S.E., Poff, N.L. and Naiman, R.J. (2006) The challenge of providing environmental flow rules to sustain river ecosystems. *Ecological Applications* 16, 1311–1318. doi:10.1890/1051-0761(2006)016[1311:TCOPEF]2.0.CO;2

Biswas, A.K. (2004) Integrated Water Resources Management: a reassessment. A Water Forum Contribution. *Water International* 29 (2), 248–256. doi:10.1080/02508060408691775

Bunn, S.E. and Arthington, A.H. (2002) Basic principles and ecological consequences of altered flow regimes for aquatic biodiversity. *Environmental Management* 30, 492–507. doi:10.1007/s00267-002-2737-0

Chéné, J.-M. (2009 Integrated water resources management: theory versus practice. *Natural Resources Forum* 33, 2–5. doi:10.1111/j.1477-8947.2009.01203.x

DWAF (1997) *White Paper on a National Water Policy for South Africa.* Department of Water Affairs and Forestry, Pretoria, South Africa. (Version) available at: www.africanwater.org/wp3.htm (accessed December 2012).

Economic Glossary (2012) Definition of allocative efficiency. Available at: http://glossary.econguru.com/economic-term/allocative+efficiency (accessed February 2013).

eFlowNet (2010) Global Environmental Flows Network. Available at: www.eflownet.org (accessed December 2012).

Esselman, P. and Opperman, J. (2010) Overcoming information limitations for developing an environmental flow prescription for a Central American River. *Ecology and Society* 15(1): 6. Available at: www.ecologyandsociety.org/vol15/iss1/art6/ (accessed December 2012).

Falkenmark, M. (2003) *Water Management and Ecosystems: Living with Change.* TEC Background Papers No. 9, Global Water Partnership Technical Committee, Stockholm.

Gordon, L.J., Finlayson, C.M. and Falkenmark, M. (2010) Managing water in agriculture for food production and other ecosystem services. *Agricultural Water Management* 94, 512–519. doi:10.1016/j.agwat.2009.03.017

GWP Eastern Africa and GWP Southern Africa (2010) *Improving Africa's Water Security: Progress in Integrated Water Resource Management in Eastern and Southern Africa.* Global Water Partnership Eastern Africa, Entebbe, Uganda and Global Water Partnership Southern Africa, Pretoria, South Africa.

GWP Technical Advisory Committee (2000) *Integrated Water Resources Management.* TEC Background Papers No. 4, Global Water Partnership Technical Committee, Stockholm.

GWP ToolBox (2008) Global Water Partnership Toolbox for Integrated Water Management. Available at: http://www.gwptoolbox.org/ (accessed February 2013).

GWSP (2013) Global Water Needs Initiative (GWNI) of the Global Water System Project (GWSP): Scientific Framework. GWSP International Project Office, Bonn, Germany. Available at: http://www.gwsp.org/scientific_framework.html (accessed February 2013).

Hughes, D.A. and Hannart, P. (2003) A desktop model used to provide an initial estimate of the ecological instream flow requirements of rivers in South Africa. *Journal of Hydrology* 270, 167–181. doi:10.1016/S0022-1694(02)00290-1

IUCN (2010) Pangani River Basin Flow Assessment. International Union for Conservation of Nature, Gland, Switzerland. Available at: www.iucn.org/about/work/programmes/water/wp_where_we_work/wp_our_work_projects/wp_our_work_pan/pangani_fa_reports/ (accessed February 2013).

IWMI (2007) The Global Environmental Flow Calculator (GEFC). Available at: http://www.iwmi.cgiar.org/Tools_And_Resources/Models_and_Software/GEFC/index.aspx (accessed February 2013).

Keys, P., Barron, J. and Lannerstad, M. (2012) *Releasing the Pressure: Water Resource Efficiencies and Gains for Ecosystem Services.* United Nations Environment Programme (UNEP), Nairobi and Stockholm Environment Institute (SEI), Stockholm.

King, J., Brown, C. and Sabet, H. (2003) A scenario-based holistic approach to environmental flow assessments for rivers. *River Research and Applications* 19, 619–639. doi:10.1002/rra.709

King, J., Turpie, J., Brown, C., Clark, B., Beuster, H. and Joubert, A. (2010) *Pangani River Basin Flow Assessment: Final Project Summary Report, December 2009.* Pangani Basin Water Board (formerly Office), Moshi, Tanzania and International Union for Conservation of Nature (IUCN) Eastern and Southern Africa Regional Program, Nairobi, Kenya. Available at: https://cmsdata.iucn.org/downloads/final_project_summary_report.pdf (accessed February 2013).

Konrad, C.P. *et al.* (2011) Large-scale flow experiments for managing river systems. *BioScience* 61, 948–959.

Korsgaard, L., Jønch-Clausen, T., Rosbjerg, D. and Schou, J.S. (2008) A service and value based approach to estimating environmental flows in IWRM. *International Journal of River Basin Management* 6, 257–266.

Kundzewicz, Z.W., Mata, L.J., Arnell, N.W., Doll, P., Kabat, P., Jimenez, B. and Shiklomanov, I.A. (2007) Fresh water resources and their management. In: Canziani, O.F., Palutikof, J.P., van der Linden,P.J., Hansen, C.E. and Parry, M.L. (eds) *Climate Change 2007: Impacts, Adaptation and Vulnerability. Contributions of Working Group II to the Fourth Assessment Report of the Intergovernmental Panel on Climate Change.* Cambridge University Press, Cambridge, UK, pp. 172–210.

Le Quesne, T., Kendy, E. and Weston, D. (2010) *The Implementation Challenge: Taking Stock of Governmental Policies to Protect and Restore Environmental Flows.* WWF Report, The Nature Conservancy, Arlington, Virginia and WWF (World Wide Fund for Nature)-UK, Godalming, UK.

Matz, M. (2008) Rethinking IWRM under cultural considerations. In: Scheumann, W., Neubert, S. and Kipping, M. (eds) *Water Politics and Development Cooperation: Local Power Plays and Global Governance.* Springer-Verlag, Berlin/Heidelberg, Germany, pp. 176–201. doi:10.1007/978-3-540-76707-7_8

Medema, W., McIntosh, B.S. and Jeffrey, P.J. (2008) From premise to practice: a critical assessment of integrated water resources management and adaptive management approaches in the water sector. *Ecology and Society* 13(2): 29. Available at: www.ecologyandsociety.org/vol13/iss2/art29/ (accessed December 2012).

Millennium Ecosystem Assessment (2005) *Ecosystems and Human Well-being: Wetlands and Water – Synthesis. A Report of the Millennium Ecosystem Assessment.* World Resources Institute, Washington, DC. Available at: www.maweb.org/documents/document.358.aspx.pdf (accessed February 2013).

Molden, D. (ed.) (2007) *Water for Food, Water for Life: Comprehensive Assessment of Water Management in Agriculture.* Earthscan, London, in association with International Water Management Institute (IWMI), Colombo.

Molden, D., Tharme, R., Abdullaev, I. and Puskur, R. (2007) Managing irrigation systems. In: Scherr, S.J. and McNeely, J.A. (eds) *Farming with Nature. The Science and Practice of Ecoagriculture*. Island Press, Washington, DC, pp. 231–249.

Naiman, R.J., Bunn, S.E., Nilsson, C., Petts, G.E., Pinay, G. and Thompson, L.C. (2002) Legitimizing fluvial ecosystems as users of water: an overview. *Environmental Management* 30, 455–467.

Nguyen-Khoa, S. and Smith, L.E.D. (2008) Fishing in the paddy fields of monsoon developing countries: re-focusing the current discourse on the 'multifunctionality of agriculture'. Keynote paper presented at the INWEPF-ICID Workshop of the RAMSAR COP10 Meeting, Changwon, Korea, 28 October–04 November 2008.

Nguyen-Khoa, S. and Smith, L. (2010) *A Cross-scale Approach to Multiple Use of Water*. Multiple Use TWG (Topic Working Group) Position Paper, CGIAR Challenge Program on Water and Food, Colombo.

Pahl-Wostl, C. (2009) A conceptual framework for analysing adaptive capacity and multi-level learning processes in resource governance regimes. *Global Environmental Change* 19, 354–365. doi:10.1016/j.gloenvcha.2009.06.001

Pahl-Wostl, C., Sendzimir, J., Jeffrey, P., Aerts, J., Berkamp, G. and Cross, K. (2007) Managing change toward adaptive water management through social learning. *Ecology and Society* 12(2): 30. Available at: www.ecologyandsociety.org/vol12/iss2/art30/ (accessed December 2012).

Poff, N.L., Allan, J.D., Bain, M.B., Karr, J.R., Prestegaard, K.L., Richter, B.D., Sparks, R.E. and Stromberg, J.C. (1997) The natural flow regime: a paradigm for river conservation and restoration. *BioScience* 47, 769–784.

Poff, N.L., Allan, J.D., Palmer, M.A., Hart, D.D., Richter, B.D., Arthington, A.H., Meyer, J.L., Stanford, J.A. and Rogers, K.H. (2003) River flows and water wars? Emerging science for environmental decision making. *Frontiers in Ecology and the Environment* 1, 298–306.

Poff, N.L. *et al.* (2010) The ecological limits of hydrologic alteration (ELOHA): a new framework for developing regional environmental flow standards. *Freshwater Biology* 55, 147–170. doi:10.1111/j.1365-2427.2009.02204.x

Postel, S. and Richter, B. (2003) *Rivers for Life: Managing Water for People and Nature*. Island Press, Washington, DC.

PRBMP (2010) Pangani River Basin Management Project: Towards Integrated Water Resource Management. Pangani River Basin Management Project (PRBMP), Moshi, Tanzania. Available at: http://www.panganibasin.com/index.php/prbmp/ (accessed February 2013).

Rahaman, M. and Varis, O. (2005) Integrated water resources management: evolution, prospects and future challenges. *Sustainability: Science, Practice, and Policy* 1, 15–21.

Richter, B.D. (2009) Re-thinking environmental flows: from allocations and reserves to sustainability boundaries. *Rivers Research and Applications* 25, 1–12.

Rowlston, W.R. and Tharme, R.E. (2008) *Determining and Implementing Environmental Water Requirements for Wetlands. Maintaining and Restoring Wetland Ecosystem Services*. Unpublished Technical Report to the Ramsar Convention on Wetlands.

Sanderson, J.S., Rowan, N., Wilding, T., Bledsoe, B.P., Miller, W.J. and Poff, N.L. (2011) Getting to scale with environmental flow assessment: the watershed flow evaluation tool. *River Research Applications* 28, 1369–1377. doi:10.1002/rra.1542

Saravanan, V.S., McDonald, G.T. and Mollinga, P.P. (2009) Critical review of integrated water resources management: moving beyond polarised discourse. *Natural Resources Forum* 33, 76–86. doi:10.1111/j.1477-8947.2009.01210.x

Seckler, D., Molden, D. and Sakthivadivel, R. (2003) The concept of efficiency in water resources management and policy. In: Kijne, J.W., Barker, R. and Molden, D. (eds) *Water Productivity in Agriculture: Limits and Opportunities for Improvement*. Comprehensive Assessment of Water Management in Agriculture Series 1. CAB International, Wallingford, UK in association with International Water Management Institute (IWMI), Colombo, pp. 37–51.

Sharma, B. *et al.* (2010) The Indus and the Ganges: river basins under extreme pressure. *Water International* 35, 493–521. doi:10.1080/02508060.2010.512996

Smakhtin, V.U. and Anputhas, M. (2006) *An Assessment of Environmental Flow Requirements of Indian River Basins*. IWMI Research Report 107, International Water Management Institute (IWMI), Colombo. doi:10.3910/2009.106

Smakhtin, V. and Eriyagama, N. (2008) Developing a software package for global desktop assessment of environmental flows. *Environmental Modelling and Software* 23, 1396–1406. doi:10.1016/j.envsoft.2008.04.002

Smakhtin, V., Revenga, C. and Döll, P. (2004) *Taking into Account Environmental Water Requirements in Global-scale Water Resources Assessments*. Research Report 2, Comprehensive Assessment of Water Management in Agriculture, Comprehensive Assessment Secretariat, Colombo. doi:10.3910/2009.391

Smith, M., and Cartin M. (2011) *Water Vision to Action: Catalyzing Change Through the IUCN Water and Nature Initiative. Results Report*. International Union for Conservation of Nature (IUCN), Gland, Switzerland.

Snellen, W.B. and Schrevel, A. (2004) *IWRM: For Sustainable Use of Water; 50 Years of International Experience with the Concept of Integrated Water Management. Background Document to the FAO/ Netherlands Conference on Water for Food and Ecosystems, 31 January–5 February 2005, October 2004*. Alterra Report 1143, Ministry of Agriculture, Nature and Food Quality, The Hague, the Netherlands and Alterra Research Institute, Wageningen University, Wageningen, the Netherlands. Available at: http://edepot.wur.nl/30428 (accessed February 2013).

Tennant, D.L. (1976) Instream flow regimens for fish, wildlife, recreation and related environmental resources. *Fisheries* 1, 6–10.

Tharme, R.E. (2003) A global perspective on environmental flow assessment: emerging trends in the development and application of environmental flow methodologies for rivers. *River Research and Application* 19, 397–441. doi:10.1002/rra.736

UN Water (2012) Draft Executive Summary of the Status Report on the Application of Integrated Approaches to Water Resources Management Draft submitted to the UNCSD (UN Commission on Sustainable Development) 2012 Preparatory Process October 28, 2011, UN Water/UNDP (United Nations Development Programme). Available at: http://www.unwater.org/downloads/UN-Water_Rio_20_report_ Draft_Exec_Sum_2011-10-28.pdf (accessed February 2013).

UNEP (2011) *Towards a Green Economy: Pathways to Sustainable Development and Poverty Eradication*. United Nations Environment Programme, Nairobi. Available at: http://www.unep.org/greeneconomy/ GreenEconomyReport/tabid/29846/language/en-US/Default.aspx (accessed February 2013).

van Koppen B., Moriarty, P. and Boelee, E. (2006) *Multiple-use Water Services to Advance the Millennium Development Goals*. Research Report 98, International Water Management Institute (IWMI), Colombo. doi:10.3910/2009.098

van Koppen, B., Smits, S., Moriarty, P., Penning de Vries, F., Mikhail, M. and Boelee, E. (2009) *Climbing the Water Ladder: Multiple-use Water Services for Poverty Reduction*. Technical Paper Series No. 52, IRC International Water and Sanitation Centre, The Hague, the Netherlands and International Water Management Institute (IWMI), Colombo. Available at: http://www.irc.nl/content/download/ 144824/465019/file/TP52_Climbing_2009.pdf (accessed February 2013).

Weber, J-L. (2011) *An Experimental Framework for Ecosystem Capital Accounting in Europe*. EEA Technical Report No. 13/2011. European Environment Agency, Copenhagen, Denmark.

WRG (2009) Charting Our Water Future: Economic Frameworks to Inform Decision-making. 2030 Water Resources Group. Available at: www.2030waterresourcesgroup.com/water_full/Charting_Our_Water_ Future_Final.pdf (accessed December 2012).

11 Management of Water and Agroecosystems in Landscapes for Sustainable Food Security

Eline Boelee,[1*] Sara J. Scherr,[2] Petina L. Pert,[3] Jennie Barron,[4] Max Finlayson,[5] Katrien Descheemaeker,[6] Jeffrey C. Milder,[2] Renate Fleiner,[7] Sophie Nguyen-Khoa,[8] Stefano Barchiesi,[9] Stuart W. Bunting,[10] Rebecca E. Tharme,[11] Elizabeth Khaka,[12] David Coates,[13] Elaine M. Solowey,[14] Gareth J. Lloyd,[15] David Molden[7] and Simon Cook[16]

[1]Water Health, Hollandsche Rading, the Netherlands; [2]EcoAgriculture Partners, Washington, DC, USA; [3]Commonwealth Scientific and Industrial Research Organisation (CSIRO), Cairns, Queensland, Australia; [4]Stockholm Environment Institute, University of York, UK and Stockholm Resilience Centre, Stockholm University, Stockholm, Sweden; [5]Institute for Land, Water and Society (ILWS), Charles Sturt University, Albury, New South Wales, Australia; [6]Plant Production Systems, Wageningen University, Wageningen, the Netherlands; [7]International Centre for Integrated Mountain Development (ICIMOD), Kathmandu, Nepal; [8]World Water Council (WWC), Marseille, France; [9]International Union for Conservation of Nature (IUCN), Global Water Programme, Gland, Switzerland; [10]Essex Sustainability Institute, University of Essex, Colchester, UK; [11]The Nature Conservancy (TNC), Buxton, UK; [12]United Nations Environment Programme (UNEP), Nairobi, Kenya; [13]Secretariat of the Convention on Biological Diversity (CBD), Montreal, Canada; [14]The Arava Institute for Environmental Studies (AIES), Hevel Eilot, Israel; [15]UNEP–DHI Centre for Water and Environment, Hørsholm, Denmark; [16]CGIAR Research Program on Water, Land and Ecosystems, Colombo, Sri Lanka

---

Abstract

Various food and financial crises have increased the pressure on natural resources while expanding on alternative ways of considering agroecosystems as potential long-term providers of ecosystem services if managed in a sustainable and equitable way. Through the study of interrelations between ecosystems, water and food security, this book has aimed to increase the understanding and knowledge of these interactions for better planning and decision making processes at various levels. This chapter concludes *Managing Water and Agroecosystems for*

* E-mail: e.boelee@waterhealth.nl

Food Security. It discusses the main findings of the preceding chapters, from analyses of drivers of sustainable food security, via agroecosystems with their ecosystem services and challenges for water use and scarcity, to specific challenges for environments such as drylands and wetlands. Using a comprehensive landscape approach, recommendations on water productivity, agroecosystem services and integrated water management are brought together succinctly. In addition, knowledge gaps and issues for further research have been identified that may support further implementation of the agroecological approach in many landscapes around the world.

Background

At the global scale, humanity is increasingly facing rapid changes, and sometimes shocks, that affect the security of our food systems and the agroecosystems that are the ultimate sources of food (Chapter 2). Together, drivers such as demographic changes, globalization of economic and governance systems (including markets), and climate change, all have important implications for the sustainable security of food. These drivers centre around food availability and access to water, as these are the major influences affecting agricultural water demand and increasing the pressure on ecosystems. Addressing the opportunities, synergies and constraints of these multiple drivers will be critical for policy advice to enable the building of resilient food systems for future generations (Chapter 2).

Water is already scarce at various temporal and spatial scales, although these estimates are hampered by uncertainties in data. However, there *is* certainty that improving food security will place further pressure on both water resources and ecosystems. How water is used in agriculture, and over time, depends on a variety of factors, including population growth, economic development, environmental constraints and accessibility, be it through infrastructure and technologies, or through governance and institutions (Chapter 5).

With increasing competition over access to water, finding an equitable way to distribute water among uses and users seems difficult, as many obstacles exist to the effective implementation of integrated water resource management (IWRM) (Chapter 10). If current water management practices continue, it is unlikely that this will solve the many challenges associated with water usage in agriculture –

challenges both from poverty and from the environment. When water for ecosystems and water for food are considered separately, the potential for inter-sectoral conflict is heightened and the problem becomes even more challenging. Hence, in order to share a scarce resource and guarantee long-term sustainability, it is imperative to transform governance theories into a practical process providing solutions on the ground that can meet future water demands.

Moving Towards an Agroecological Landscape Approach

Chapter 6 discusses several approaches that are capable of preserving water and increasing the use of low quality water for dryland agriculture. Drylands are highly vulnerable, making ecosystem management a priority for reversing land degradation and making optimum use of the – often limited – available natural resources. Options such as the introduction of appropriate plant species for specific landscapes can be successful only when the entire ecosystem and its users are taken into account. Similarly, the exploitation of wetlands by people is a reality that has already led to rapid degradation. Wetlands sustain a wide range of ecosystem services that contribute to water and food security, but their exploitation should be embedded in a systematic approach that incorporates the many functions of these ecosystems, or more will be lost and further degraded (Chapter 7).

Management of water in agriculture typically targets the provisioning ecosystem services of biomass for the harvesting of food, fodder, fibre or other valued goods, often at the expense of other supporting or regulatory services (Chapter 3). Reduced access to and quality of water, combined with increased demand for water, are among the reasons for the rapid growth of aquaculture, which has its own water management requirements and impacts on water and food security. At the same time, water availability is one of the limiting factors to biomass production, which explains why agricultural management is often focused primarily around the supply of water, in combination with other inputs such as manure, fertilizer, improved seeds and pest control (Chapter 5). If agricultural activities are viewed in isolation and receive disproportionately more water, landscapes will lose the capacity to provide the full range of ecosystem functions and services that they currently do – or formerly did (FAO/Netherlands, 2005). However, as long as many of these ecosystem services do not have a market value, they are not included in conventional agricultural management approaches. At the same time, conservation strategies usually target specific (semi-natural) ecosystems or threatened species, while treating agriculture as a threat to be contained and mitigated. In Chapter 4, it was shown that it is possible to go beyond making trade-offs between agricultural production and environmental quality by improving the quality of agroecosystems and reversing degradation that has an impact on productivity.

Ecosystem services are directly important for many people (Chapter 3). This is particularly the case in agriculture-based basins in low income countries, where many livelihoods depend on natural ecosystems such as wetlands and forests for survival (Kemp-Benedict et al., 2011). Pro-poor policy responses to environmental problems can enhance multiple objectives such as human health, socio-economic growth and aquatic environmental sustainability (Millennium Ecosystem Assessment, 2005). Healthy agroecosystems have the potential to provide a high diversity of nutritious food, which is based on their functioning biodiversity (Kaplan et al., 2009). Sustainable management of agroecosystems is, therefore, critical to addressing food security issues (FAO/Netherlands, 2005).

To ensure food security, it is important that decision makers support the management of ecosystem services by taking appropriate policy measures that encourage sustainable land management, integrated water resources management and more sustainable agricultural practices by farmers. Solutions include not only minimizing the negative impact of agriculture on ecosystems, but also better management of agroecosystems and non-agricultural ecosystems to support improved water security for agriculture. These approaches include, among others, equitable access rights, better soil and water conservation, management of water quality and quantity, improved livestock and fish management, and sustaining biodiversity (Chapter 9).

This calls for a refining in the management of water from what is termed in IWRM 'water for food' to 'water for agroecosystems', in which the whole ecosystem base of provisioning, regulatory, cultural and supporting services are considered (Chapter 10). Recognizing and accounting for these multiple ecosystem services of agroecosystems, coupled with elements of IWRM at the basin scale, including consideration of all water resources above and below ground, can be a powerful and sustainable response to freshwater scarcity. Identifying the most promising options to increase water productivity is complex and has to take into account environmental, financial, social and health-related factors. In general, improving agricultural water productivity can be achieved by creating synergies across scales and between various agricultural sectors and ecosystems, thereby enabling multiple uses of water and equitable access to water resources for different groups in society (Chapter 8).

Lessons Learned: Principles and Recommendations

Many of the recommendations in the previous chapters may prove most effective if they are applied in combination with other measures in an integrated approach, as has been shown for wetlands (Chapter 7). Increased water productivity in agroecosystems enhances the value of water within those systems and helps in the reallocation of water to a variety of uses and ecosystems (Chapter 8). In multifunctional agroecosystems, water is thus managed more productively for provisioning services (crops, trees, livestock, fish) and, in addition, sustains regulating, supporting and cultural ecosystem services (Chapter 5). In an IWRM approach, this can be managed at the basin level for agroecosystems, cities, industry, nature and other functions (Chapter 10). In many cases, active management of the landscape and various elements in it is required to help sustain the various ecosystem functions and services (Molden *et al.*, 2007).

In an agroecological landscape approach, ecosystems are linked, and natural resources, such as water and land, are managed specifically to enhance ecosystem services. In this way, synergies can be exploited and productivity improved, while obtaining added value from improved carbon storage, erosion control, water retention, waste treatment and cultural values such as recreation (Chapter 4). As pointed out in the previous chapters, most of these added services do not conflict with agricultural production and, in many cases, they improve both its productivity and sustainability, so that the integrated management of agroecosystems for multiple services and benefits can be considered to be the key for addressing food security issues (FAO/ Netherlands, 2005). The challenge is then to manage agroecosystems and landscape ecosystem services so that that improved water management and increased water productivity lead to both food security and synergies within the landscape, instead of to mutual degradation (Keys *et al.*, 2012).

Based on the findings reported in this book for drylands (Chapter 6), wetlands (Chapter 7), increased water productivity (Chapter 8), the management of ecosystem services (Chapter 9) and integrated water management (Chapter 10), and supplemented with references from the international literature, we have summarized a set of recommendations below on how to manage water and other natural resources in agroecological landscapes. While basins are the ideal management unit from a hydrological or IWRM perspective, in reality, administrative boundaries play an important role and a more flexible definition of the landscape has to be used.

- *Prioritize development issues*
  Each landscape or basin, depending on the context, has its own issues. Long-term problems may be quite similar, but short-term priorities need to be determined locally. The analyses by Cook *et al.* (2011) and Kemp-Benedict *et al.* (2011) provide a useful starting point by considering the level of economic development within a particular locality or region (discussed in Chapter 5). The process of formulating priorities can further be facilitated by using some of the guiding questions suggested by Cook *et al.* (2011). For instance: how much water is there in the basin and who uses the water; how productively is water used by agriculture; and who has the power to change this? Finally, interventions can be developed, for instance, by looking at how sensitive the system is to change (see Cumming, 2011).

- *Promote diversity within the production systems*
  Multifunctional agriculture can help to increase the productivity of natural resources and reduce risk for farmers (OECD, 2001; Groenfeldt, 2006). Optimizing the diversity of the above and below ground biotic components within the production system (crops, animals, soil and pollinators) can increase the adaptive capacity of the production system. This would help to buffer it against fluctuations in water availability, temperature, and pests and diseases, thereby enhancing the resilience of the system as well as rural livelihoods (Hajjar *et al.*, 2008; Chapter 9). Synergies between livestock and aquaculture (van der Zijpp *et al.*, 2007) can be explored for increasing resource recovery and productivity (Chapter 8). The same holds true for other integrated systems, such as crop–livestock systems, rice–fish culture, tree–crop systems (Zomer *et al.*, 2009), aquaculture in reservoirs, forest pastures or wastewater-fed aquaculture (Chapter 7). The integration of trees can help to fix nitrogen, tighten nutrient, water and carbon cycles, and produce

additional goods, e.g. year-round available fodder, and biomass for use as organic fertilizer and fuel (Garrity et al., 2010).

- *Promote diversity in landscapes*

Landscapes with high levels of land use diversity, as well as biodiversity, are more resilient and better able to mitigate adverse environmental conditions (Folke et al., 2004). Large monocropped areas can be developed into landscapes that have higher levels of biodiversity by identifying and linking natural habitat patches, including aquatic ecosystems. Habitat integrity and connectivity can be maintained by incorporating hedgerows, multipurpose trees and, where spatially feasible, corridors of natural vegetation interconnecting parcels of agricultural land and natural ecosystems (such as wetlands and forests, which may need to be specifically developed where these natural systems are remote). In large irrigated areas, canals and roads can be lined with perennial vegetation, such as trees, thus also serving as important passages and habitats for animals. Canals and other waterways that connect aquatic ecosystems, and so maintain the connectivity of migratory routes for aquatic fauna, provide the variety in habitats required for subsequent life cycle stages, for example spawning (Chapter 7). Landscape-scale planning of strategic tree cover interventions can reduce flow accumulation by providing sites for water infiltration and penetration. For instance, contour hedgerows can reduce runoff and soil erosion on slopes, and buffer strips may protect watercourses (Chapter 9). By incorporating both fodder production and grazing land, livestock can be managed at the landscape level too, so that animals are enabled to reach otherwise inaccessible feed sources and the overgrazing and trampling of vulnerable areas is avoided.

- *Increase water productivity*

Water productivity is defined as the amount of output per unit of water. Crop water productivity can be improved by selecting well-adapted crop types, reducing unproductive water losses and optimizing the joint management of water, nutrients and plants. However, it is crucial to go beyond crops, and to include livestock, trees and fish in water productivity assessments (Chapter 8). Livestock water productivity can be increased through improved feed and animal management, reducing animal mortality, appropriate livestock watering and sustainable grazing management. In agroforestry systems, the right combination of trees and crops can exploit spatial and temporal complementarities in resource use. In aquaculture systems, most water is depleted for feed production, via seepage and evaporation, and through polluted water discharge. Hence, efforts to minimize those losses would improve overall water productivity.

- *Choose the right infrastructure and operation*

Smart infrastructure planning, and selecting appropriate, multifunctional constructs at the right location that can be operated with a large degree of flexibility, can widen the focus, from simply delivering water to field crops, to providing water for multiple uses by different members of society. This would explicitly include water for bathing, laundry, animal drinking, home gardens, fish ponds and many other domestic and productive uses. Current access to water has to be taken into account and, where necessary, expanded with appropriate structures for the harvesting of rain or runoff water, site-specific water storage (McCartney and Smakhtin, 2010) and distribution infrastructure, as well as by using unconventional sources of water (such as urban wastewater, which can be a valuable resource if managed properly). This approach would need to take into account property rights and their gendered nature, including the rights to the use and management of shared water resources.

- *Mobilize social organization and collective action*

Engaging communities in water resource management and ownership is critical to ensuring that the various proposed practices meet the needs of the people and are carried on into the future for meeting food and environmental needs. This includes management of other natural resources, such as land and common forest, and grazing lands (Bossio and Geheb, 2008). Alternative grazing management practices can have a substantial impact only when compliance is high, and sufficient spatial coverage of the interventions is ensured. Efforts also need to be made to make sure that management and ownership involve equitable access across diverse and sometimes marginalized groups within local communities. The devolution of responsibilities has to be matched with the devolution of rights or power. Raising awareness among community groups about the implications of alternative types of water use and the associated trade-offs will greatly enhance the capacity of the groups to conserve biodiversity and manage water efficiently.

• *Apply refined IWRM*
In IWRM, all sources of water throughout the basin, including rain and surface water, as well as water held in soil and aquifers, are considered in a comprehensive manner and managed for the benefit of a broad range of uses. As stated in Chapter 10, water is no longer supplied to crops, trees, livestock or fish, but to multifunctional agroecosystems linked and managed together, at the river basin or landscape level. With effective institutional and policy support, water use can be optimized by increasing its efficiency, for example by using water more effectively in rainfed agriculture, improving water storage and reusing wastewater, as well as limiting the further expansion of water withdrawal from water sources and minimizing the impacts of climate change. IWRM can be further developed by recognizing and incorporating the need to safeguard environmental water flows for other ecosystems in order to enhance long-term ecosystem health and resiliency. The concept of environmental flows provides a basis for calculating the amount of water (quantity, quality and timing) required to sustain ecosystems and safeguard their services to people (Keys *et al.*, 2012; Chapters 5 and 10).

• *Develop institutions for integrated natural resources management*
Up until now, relatively more effort has been placed on building institutions to manage irrigation delivery than on overall water and natural resource management. However, institutions must be developed, changed and supported at various levels to maintain healthy multifunctional agroecosystems and to ensure equity of access, use and control over resources. This means that specialized line agencies from various ministries have to collaborate much more closely than before, both with each other and with the end users. For example, joint management of water, land, crop and livestock is required to adequately address erosion problems.

• *Foster supportive policies*
National and landscape level policies can support not only the development and management of early warning and response systems for climate change, but also improved markets, buffers of food and fodder, and insurance schemes to cover loss of yields or livestock (World Bank, 2009). Multifunctional agroecological landscapes need supporting services in all sectors, ranging from the monitoring of water distribution and soil fertility to veterinary centres and public health facilities. Incentive systems such as payments for ecosystem services (PES) and payments for watershed services can support the transition to more sustainable farming systems and enhance resilience (Swallow and Meinzen-Dick, 2009; Mulligan *et al.*, 2010; Cumming, 2011; Chapter 9). Such policies may enable farmers to adopt practices that lead to long-term benefits, but with lower returns in the short term.

• *Inter-sectoral collaboration*
The application of an integrated approach to water resources management, and the strengthening of institutions and policies for an ecological landscape approach, require an enlightened collaboration across different relevant sectors (Chapter 5), e.g. between the various ministries, both at a national and a local level, depending on the level of decentralization in a country. Ministries of the Environment may be in the best position to promote an ecosystem services approach to food security at the landscape level, because of their expertise, but, in reality, this is a huge challenge. In

many countries Ministries of the Environment, if they exist at all, often have a small budget and little power. In contrast, Ministries of Agriculture are often in a better position, especially in countries where agriculture delivers a large proportion of the gross domestic product (GDP). Also, the mandate, policies and practices in the agriculture sector probably have the greatest potential to help shape agroecological landscapes. Therefore, if all the various ministries work together, with the water, health, energy and planning sectors, there would be greater potential leverage for promoting the management of agroecosystem services.

The Integrated Landscape Approach in Practice

Landscape approaches can improve food security and nutrition by diversifying food sources and increasing the sustainability of production systems at multiple scales. It takes the concept of multifunctionality in agriculture (Nguyen-Khoa and Smith, 2008) to a higher scale. For example, multifunctional rice fields in Vietnam are used to grow rice (thus increasing food security), reduce erosion and buffer water quantities (both regulatory services), retain nutrients (supporting services) and, at the same time, diversify production by allowing fish and ducks into the rice fields, as well as grazing animals after harvest. Landscape elements can be added to enhance more regulatory and supporting ecosystem services, e.g. ponds interspersed with the fields, which can be used for aquaculture and livestock drinking, while also regulating water flows. Similarly, multipurpose trees can help to increase infiltration and reduce runoff (regulatory services), and can be used in agricultural landscapes to connect forest habitats (biodiversity), bring insects for pollination and soil organisms closer to fields (supporting services), cycle nutrients and carbon (supporting services), and also diversify provisioning services by supplying fuel wood and timber in addition to fodder and fruit.

In aquatic ecosystems, recognition of the full range of provisioning ecosystem services,

i.e. not only fish, is vital if the true value of such 'aqua-ecosystems' is to be accounted for and safeguarded in the livelihoods of people and in local and national economies. Beyond capture fisheries, well-managed aquatic ecosystems also provide biodiversity, cultural services and aesthetic values. By providing environmental flows at the basin level, aquatic ecosystem services may benefit many people and make significant contributions to their well-being and resilience (Brummett et al., 2010).

When managing agroecosystems as part of landscape approaches, the upstream areas merit special attention, as these are often degraded and need to be rehabilitated. In practice, many of the examples, particularly those relating to action in wetlands or changing agricultural practices, are about ecosystem restoration (e.g. Box 11.1). In most cases, this means restoring or enhancing the services provided by the ecosystem. This also holds true for the transformation of conventionally managed agroecosystems into multifunctional agroecological landscapes that provide the widest range of ecosystem services. Rehabilitation implies the regrowth of grass and trees, which ultimately requires water. Hence, in these areas, the water for agroecosystems will not immediately result in many provisioning services, except possibly for some fodder as part of cut and carry systems. During this rehabilitation phase then, there may be higher water requirements for regulatory and supporting services (including carbon capture), with less water to contribute to downstream river flows. In the long term, there will be compensation for this phase through a reduction in erosion levels (and in the siltation of downstream infrastructure), and increased infiltration and higher downstream river flows.

Several organizations are promoting landscape approaches and contribute to a knowledge base on the impacts and constraints. The 'Landscapes for People, Food and Nature Initiative' (2012) identified 109 active or recent landscape initiatives in Latin America, 150 in Africa and 21 in Asia and the Middle East. In these landscape management systems, people have developed integrated strategies and multi-stakeholder processes for maintaining agricultural productivity, as well as rich natural ecosystems. Many of these are projects in

---

Box 11.1. River restoration in Jordan. Example based on IdRC (2006), IUCN (2009), and updated by Stefano Barchiesi (IUCN) in April 2012.

The Zarqa River is the second tributary to the Jordan River. It rises in springs near Amman and flows through a deep and broad valley into the Jordan River. Around 65% of the Jordanian total population and more than 90% of the small–medium industries of Jordan are concentrated in the Zarqa River Basin, and the demands for water are very high. This has led to the over-pumping of groundwater for agriculture, drinking and industrial uses, which, together, have reduced the natural base flow of the river. The flow characteristics have been further modified by the discharge of treated domestic and industrial wastewater into the river; these discharges comprise nearly all of the summer flow of the river and substantially degrade the water quality.

In a heavily populated and industrialized region, it is a challenge to establish a solid waste management strategy to stop the contamination of the river. Since 2006, the REWARD (Regional Water Resources and Drylands) Programme of the IUCN (International Union for the Conservation of Nature) Regional Office for West Asia (ROWA) has worked in close consultation with the Jordan Ministry of Environment on a long-term strategy for the restoration of the basin. 'The Restoration and Economic Development of Zarqa River Basin' Project was one of the demonstration projects initiated by the DGCS (Italian Ministry of Foreign Affairs)/WESCANA (West, Central Asia and North Africa) Project and supported by the IUCN Water and Nature Initiative (WANI), with funding from DGIS (Netherlands' Directorate-General for International Cooperation).

The Jordanian Ministry of Environment placed the rehabilitation and integrated environmental management of the Zarqa River Basin at the top of its priorities in 2006. With the support of IUCN, the Ministry conducted a sustainability review of the institutional arrangements within the Ministry of Environment, including the Zarqa River Basin Rehabilitation Unit (ZRRU), and formed a Committee with representatives of governmental institutions, research organizations, universities and local NGOs (non-governmental organizations) to develop a national strategy for the restoration of the Zarqa River. The strategy for action was completed in 2009 with funding from the Spanish Agency for International Cooperation and UNDP (United Nations Development Programme). The strategy builds on the principles of integrated water resources management (IWRM) in combining the development of effective governance, the application of economic tools, knowledge management and capacity building, the engagement of civil society and the implementation of restoration and sustainable management.

The restoration strategy has three phases. In the first 3 year phase, urgent pilot restoration activities demonstrated to people how progress can be achieved, and what the benefits of a healthy river are to society. At the same time, planning took place for cleaning up the rubbish in the river, re-establishing riverside vegetation and managing water resources sustainably. This was backed by the participation of river users and communities in decision making and action for rehabilitation. The economic benefits from restoration will grow over time, together with the regulatory, provisioning (agriculture) and cultural (recreation and tourism) ecosystem services. In the second and third phases, the whole river ecosystem will be restored to health over a period of 10–15 years.

The significant obstacles that remain to improving water management in the Zarqa River Basin are the lack of clear governance of the water basin, which has resulted in more difficult enforcement of water legislations and policies, and also the lack of inter-stakeholder agreements, knowledge of socio-economic consequences, information on various aspects of the basin (e.g. environmental flows and surface–groundwater interactions), awareness of water conservation and management, and experience of systematic approaches relevant to water resources management.

---

places where high levels of poverty coincide with critical conservation priorities (Landscapes for People, Food and Nature Initiative, 2012). This confirms the dependency of many poor people on ecosystems for their livelihoods (Chapter 3). In other areas of the world, whole landscape approaches are being developed particularly to address issues of water quality, water conflicts over natural resources and cultural heritage (Landscapes for People, Food and Nature Initiative, 2012). Other initiatives operate at a somewhat smaller scale but have important landscape components, such as the global initiative to identify and safeguard Globally Important Agricultural Heritage Systems and their associated landscapes, agricultural biodiversity, knowledge systems and cultures, which was started by FAO (Food and Agriculture Organization of the United Nations) in 2002 (http://www.giahs.org).

Research is now underway to track and analyse impacts on production, ecosystem services and human well-being of the landscape approach in many locations. Though methodologies are still under development, it has become clear that in many places the approach has not yet moved, beyond a vision of leaders, into practical implementation. This highlights a particular risk in adopting a landscape approach, as the notion of what a landscape should sustain and how it should be managed will vary between stakeholders. Consequently, policy makers and natural resource managers will need integrated landscape assessment and participatory planning approaches to reconcile multiple demands, including the need for environmental protection and biodiversity conservation. Concerns that landscape approaches might be dominated or corrupted by powerful or more influential groups should also be acknowledged, and appropriate safeguards and measures are needed to ensure transparency and accountability. On the optimistic side, the available case studies demonstrate that there have been ample positive results in vulnerable regions, where earlier conventional approaches did not work (see Landscapes for People, Food and Nature Initiative, 2012; Landscapes for People, Food and Nature Initiative Blog, 2012). The results that have been obtained suggest that 'landscape approaches can increase the "total bottom line" outcomes of rural landscapes while improving the sustainability of livelihoods and resilience of rural communities' (Landscapes for People, Food and Nature Initiative, 2012).

Related initiatives

Another concept that is gaining worldwide momentum is that of Green Economy. Although this is criticized for its practical implications, several recommendations on water management echo the recommendations outlined above (UNEP, 2011; MLTM *et al.*, 2012). Investment in water-dependent ecosystems, infrastructure and management is seen as a way to expedite the transition to a green economy and to help achieve all water-related Millennium Development Goals (MDGs) at the same time as keeping global water use within sustainable limits. This would require an investment of US$198 billion/year until 2050. Institutional arrangements, such as PES, could help to reduce this amount. The improvement of agricultural water use is crucial to achieving these goals (UNEP, 2011).

The British programme 'Ecosystem Services for Poverty Alleviation' (ESPA, 2009–2016; see NERC, 2012) strives to 'improve ecosystems management policies to help alleviate poverty in the developing world'. Even though it is not directly focused on food security, ESPA may help to bridge some of the knowledge gaps identified in this book by supporting 'high quality and cutting-edge research that will deliver improved understanding of how ecosystems function, the services they provide, the full value of these services, and their potential role in achieving sustainable poverty reduction'. The evidence and tools generated in ESPA should enable farmers and decision makers to manage ecosystems sustainably and in a way that contributes to reducing poverty, hunger and disease. The programme seeks to provide evidence on the values of ecosystem services, drivers and trends, processes that influence ecosystem services, the importance of these services in alleviating poverty and enhancing sustainable growth, and ways to overcome constraints for the provision of those critical ecosystem services. ESPA works in South Asia, China, sub-Saharan Africa and Amazonia, and addresses various research themes, including water, health and biodiversity.

The CGIAR Research Program on Water, Land and Ecosystems was launched in March 2012 and aims to address some of the world's most pressing problems related to boosting food production and improving livelihoods, as well as simultaneously protecting the environment. This 10 year Program capitalizes on available knowledge and solutions in natural resource management, and aims to bring these together under an umbrella of ecosystem management (WLE, 2012).

Finally, trans-disciplinary thinking is also a key component of two integrative approaches to public health: EcoHealth (Lebel, 2003) and One Health (Kaplan *et al.*, 2009). Both of these require scientists from very different disciplines to work together, and emphasize the role of a healthy environment in determining the health of people and animals (see also Chapter 5). The link with natural resource management for food security in agroecological landscapes is apparent from the many pandemics in recent years, such as H1N1 flu and avian flu, which were related to livestock and poultry management.

Implications and Priorities for Research

The gap between theory and practice highlighted in the previous section shows that there are many challenges to the effective implementation of integrated landscape approaches. Some of these are institutional constraints, as pointed out in the recommendations that we have listed, but there are also important knowledge gaps, as laid out in the preceding chapters. Briefly, while understanding and recognizing the principles, do we really know *exactly how* water and agroecosystems must be managed to achieve food security, now, in 2050 and beyond?

Knowing the various agroecosystem functions and services that water can provide, how can water management enhance these functions and services? How do we decide how much water should be used for crop irrigation and energy production, and how much should be used for nature conservation? Can we really manage our agroecosystems differently so that we reduce trade-offs, as well as maximize the provision of ecosystem services, and at the same time use the same amount of natural resources, especially of land and water? Will this indeed lead to long-term sustainability and increased well-being for more people? How can recommendations for improvements from field to landscape or basin level be translated into policy actions? Questions like these can help identify knowledge gaps, along the lines of the guiding questions formulated by Herrero *et al.* (2009) on livestock, ecosystems and livelihoods.

Knowledge gaps from preceding chapters

To ensure that we have enough water for food and for a healthy planet, we must go beyond implementing the known improved techniques, incentives and institutions, and invest in understanding the various ecosystem functions and services, as well as their interactions, in agroecosystems (Chapter 3). Much is known about the global drivers of food security (Chapter 2), but we know less about how these drivers – either directly or indirectly – affect ecosystem services. A more in-depth analysis of one of these drivers, e.g. climate change, on productivity, ecosystem services and livelihoods in nine basins led to general recommendations, but was hampered by the usual uncertainties in the predictions (Mulligan *et al.*, 2011). Even at local level, there is not always sufficient information on the value of ecosystem services, especially when these go beyond the provision of food and fuel, which makes them less suitable for monetization (Chapter 4). Recent inventories of the global values of ecosystem services, such as those by the global initiative Economics of Ecosystems and Biodiversity (TEEB, 2010), have demonstrated the huge knowledge gap that there is of the value of the agroecosystems (van der Ploeg *et al.*, 2010) that cover so much of the earth's surface, compared with other ecosystems.

Our knowledge on water use in agroecosystems is more detailed, although this may not be true for a broader range of ecosystem services (Chapters 5 and 8). More research is needed on tools to analyse the potential for improvement at various spatial and temporal scales in order to focus and tune an appropriate and practical management approach year after year. Some major areas where attention needs to be given are: the role of agroecosystems in water storage and supply, particularly the renewable recharge of groundwater and improved soil moisture storage, and the role of water transpiration by agricultural crops in sustaining local and regional water cycles (Chapter 5). Insights are limited on water quality at basin level, and its influences from and impacts on agricultural development, though this is likely to become an increasingly more important topic.

The discussion of drylands (Chapter 6) and wetlands (Chapter 7) have provided concrete examples of how complex the issues in these ecosystems can be. As the specific interventions are, by definition, site specific, general knowledge gaps here are mainly on which institutional arrangements would best support integrated ecosystem management approaches. The application of various interventions is not obvious either. For instance, many studies have suggested there is substantial potential in semi-arid rangelands for carbon sequestration as well as for increases in water productivity, but much less is known about the way these would have to be managed in order to tap in to this potential (Chapter 6). In arid and saline areas, sustainable agriculture is only possible by combining specific crop selection and crop development, soil management, integrated livestock management, conservation and agroforestry, and mobilizing underutilized water sources.

The synergy of such a combined strategy may increase the efficient use of the resource base, but it requires local and generic knowledge to be applied, with attention given to the many marginal groups that live in these areas. In wetland agroecosystems, approaches to the multiple use of water resources show great promise for increased productivity, but rely on the preservation of traditional knowledge on integrated systems and the constraints of these systems (Chapter 7). Opportunities to further develop integrated systems could be explored, though assessments are required to determine the impact on ecosystem services within and outside these systems, and the potential for new risks to farmers. Such studies will also help to identify the needs for capacity building of water and land managers on ecosystem services.

There have been many recent innovations in the field of increasing water productivity (Chapter 8), but less is known about the impacts on overall basin water efficiencies if several measures are applied at a large scale. Scenario analysis can help in understanding the effects of different options on future water demands from agriculture, but the inclusion of other sectors, such as livestock, fisheries, aquaculture and trees, as well as non-provisioning ecosystem services, is much more complicated. Hence, further research is needed on the implications of various (integrated) interventions and of improved agricultural water productivity on poverty, food security, economic growth and landscape functioning, in addition to cost–benefit analyses of water requirements for ecosystem services. At the same time, not enough is known about how innovations change our systems and the resulting outputs and impacts of the agroecosystems, but this type of research is rare. Therefore, predictions on which innovations to develop how, when and where can hardly be made. Similarly, while payment for (environmental or) ecosystems services (PES) is being applied more widely, knowledge is still lacking on the proper institutional arrangements to ensure that farmers receive sufficient incentives for their changed practices (Chapter 9).

Fundamental research questions emerge on how to operationalize the various recommendations made, e.g. how to ensure that water in IWRM is allocated to a broad range of agroecosystem services, and not just to food production (Chapter 10). Ecoagricultural research from an IWRM perspective would help to identify feasible options that, in turn, need policy efforts to be realized. Although the political will may not always be available, this could change, once countries consider the full social, economic and environmental costs of not conserving existing water resources, as well as the costs of failure to develop new water sources. Multi-stakeholder dialogues would help to generate support at all levels, but need to be based on specific knowledge, packaged and communicated in a way that is understandable to the broader population.

Other scientists have also pointed out the many knowledge gaps on how local solutions should be combined into an overall integrated landscape approach, in which issues such as competition between multiple functions and the earlier mentioned methodological issues in the valuation of ecosystem services for current and future generations are important components (Hermann et al., 2011). However, it is not always clear what policies are best, especially when evaluating decisions about ecosystems for water and food production, as these different systems operate on different timescales, but also underlie each other. Specific social science research is needed to include a better understanding of adaptive management institutions, but also of the role of women and youth and resilient livelihoods in implementing landscape approaches.

As climate variability increases across agroecological zones (Chapter 2), farmers need a reliable water supply, supported by adequate storage systems, to secure sustainable food production (Chapter 5). Assessments of freshwater scarcity continue to be refined from previous annual estimates to reflect the actual monthly consumptive use of groundwater and surface water. Still more rigorous and realistic accounting of the flows is needed to sustain the ecological integrity and essential services of those agroecosystems being managed for long-term food security, which will help to add a long-term environmental understanding to these water use patterns. Timely provision of good quality inputs is no longer sufficient, and early warning systems need to be developed to help address the vulnerability of farmers in variable climates and enhance efficiency in the use of water and nutrients (Chapter 8). The timescale needs to take into account the potential impacts of climate change, especially for vulnerable populations with low adaptive capacity, in order to ensure that food security targets are met.

Need for action research

The next step is to put these guidelines into practice and monitor the process long term and closely, to see how it works in reality and where adaptations are required in the approach. Baseline valuation assessments of ecosystem services in agricultural production systems can then be compared with those in real life agroecological landscapes, providing an evidence base on what works and what does not in the application of an ecosystem services approach to water and food security. In many places, such evaluations are hampered by an almost total lack of baseline data (see above), e.g. on streamflows and water quality, and sometimes even on rainfall. This makes it very hard to run models and obtain a good

understanding of the current situation, let alone monitor and evaluate change.

The Landscapes for People, Food and Nature Initiative (2012) has started supporting the large scale application of landscape approaches, including improving key aspects of the enabling environment. This and other initiatives (e.g. Vital Signs, 2012) are also working to improve monitoring and impact assessment of landscape level initiatives and processes, but much remains to be done. The end users need support in managing their landscapes with the natural resources in it, for a wide range of ecosystem services, with proper valuation and monitoring to determine impact, identify obstacles and successes, and develop recommendations for further practical application of the approach. This is only possible with long-term collaborative commitments and agreements between universities, research centres, landscape initiatives, planners and practitioners. While increased environmental sustainability and productivity can be expected when managing for a broad range of ecosystem services, targeted interventions are needed to safeguard food security and equity. Hence, we end this book with an appeal to collaborative action research, in which planners and implementers from the agricultural, environmental and other sectors collaborate with scientists to create and scientifically monitor agroecological landscapes over the long term.

Conclusions

This book has shown the importance of ecosystem services in agriculture and how water and ecosystems can contribute to food security. The capacity and productivity of agroecosystems will be enhanced when the water quantity and quality are adequate for the whole range of ecosystem services, which will lead to greater environmental sustainability. Summarized in this chapter are the main elements required for an integrated management approach to water and agroecosystems at the landscape level. The resulting landscapes look like a mosaic of various healthy agroecosystems, natural ecosystems and other landscape elements, in which institutions and policy effectively facilitate the supporting role of water. Thus, agroecosystems can remain productive in the long term, resulting in higher economic returns for farmers. With more ecosystem services and, therefore, greater productivity at basin or landscape level, the health of the ecosystems comprising the resource base would be enhanced, and contribute to long-term sustainable food security.

In the previous chapters, various knowledge gaps were shown, particularly in the under- standing of ecosystem services in agriculture and the implementation of the ecosystem approach. More insights can be gained through in-depth analysis of scientific as well as indigenous and practical evidence in various disciplines, which are, as yet, hardly ever combined.

By focusing on the benefits that we can derive from integrated landscape approaches for managing agroecosystems, we can shift the focus from production activities in isolation of other ecosystem services to a focus that treats our productive landscapes as settings for human well-being – settings that are main- tained through integration of the benefits that we derive from our water, food and environ- ment.

References

Bossio, D. and Geheb, K. (eds) (2008) *Conserving Land, Protecting Water.* Comprehensive Assessment of Water Management in Agriculture Series 6. CAB International, Wallingford, UK in association with CGIAR Challenge Program on Water and Food, Colombo and International Water Management Institute (IWMI), Colombo.

Brummett, R.E., Lemoalle, J. and Beveridge, M.C.M. (2010) Can water productivity metrics guide allocation of freshwater to inland fisheries? Knowledge and Management of Aquatic Ecosystems 399, 1–7. doi:10.1051/kmae/2010026

Cook, S., Fisher, M., Tiemann, T. and Vidal, A. (2011) Water, food and poverty: global- and basin-scale analysis. *Water International* 36, 1–16. doi:10.1080/02508060.2011.541018

Cumming, G.S. (2011) The resilience of big river basins. *Water International* 36, 63–95. doi:10.1080/02508 060.2011.541016

FAO/Netherlands (2005) *Report: FAO/Netherlands International Conference. Water for Food and Ecosystems. Make it Happen! The Hague, January 31–February 4, 2005.* Available at: http://www.fao.org/ag/wfe2005/docs/initialdocument.doc (accessed February 2013).

Folke, C., Carpenter, S., Walker, B., Scheffer, M., Elmqvist, T., Gunderson, L. and Holling, C.S. (2004) Regime shifts, resilience, and biodiversity in ecosystem management. *Annual Review of Ecology, Evolution, and Systematics* 35, 557–581.

Garrity, D.P., Akinnifesi, F.K., Ajayi, O.C., Weldesemayat, S.G., Mowo, J.G., Kalinganire, A., Larwanou, M. and Bayala, J. (2010) Evergreen agriculture: a robust approach to food security in Africa. *Food Security* 2, 197–214. doi:10.1007/s12571-010-0070-7

GIAHS (2002) Globally Important Agricultural Heritage Systems. GIAHS Secretariat, Food and Agriculture Organization of the United Nations, Rome. Available at: http://www.giahs.org (accessed February 2013).

Groenfeldt, D. (2006) Multifunctionality of agriculture water: looking beyond food production and ecosystem services. Prepared for the FAO/Netherlands International Conference on Water for Food and Ecosystems, The Hague, January 31–February 5, 2005. *Irrigation and Drainage* 55, 73–83.

Hajjar, R., Jarvis, D.I. and Gemmill-Herren, B. (2008) The utility of crop genetic diversity in maintaining ecosystem services. *Agriculture, Ecosystems and Environment* 123, 261–270. doi:10.1016/j.agee.2007.08.003

Hermann, A., Schleifer, S. and Wrbka, T. (2011) The concept of ecosystem services regarding landscape research: a review. *Living Reviews in Landscape Research* 5, 1. Available at: www.livingreviews.org/lrlr-2011-1 (accessed February 2012).

Herrero, M., Thornton, P.K., Gerber, P. and Reid, R.S. (2009) Livestock, livelihoods and the environment: understanding the trade-offs. *Current Opinion in Environmental Sustainability* 1, 111–120. doi:10.1016/j.cosust.2009.10.003

IdRC (2006) *The Integrated Environmental Management of the Zarqa River. Final Report.* Interdisciplinary Research Consultants, Amman, Jordan.

IUCN (2009) The Restoration and Economic Development of Zarqa River Basin. International Union for the Conservation of Nature. PowerPoint Presentation. Available at: http://cmsdata.iucn.org/downloads/pps_zarqa.ppt (accessed June 2011).

Kaplan, B., Kahn, L.H. and Monath, T.P. (2009) The brewing storm. *Veterinaria Italiana* 45, 9–18.

Kemp-Benedict, E., Cook, S., Allen, S.L., Vosti, S., Lemoalle, J., Giordano, M., Ward, J. and Kaczan, D. (2011) Connections between poverty, water and agriculture: evidence from 10 river basins. *Water International* 36, 125–140. doi:10.1080/02508060.2011.541015

Keys, P., Barron, J. and Lannerstad, M. (2012) *Releasing the Pressure: Water Resource Efficiencies and Gains for Ecosystem Services.* United Nations Environment Programme (UNEP), Nairobi and Stockholm Environment Institute (SEI), Stockholm.

Landscapes for People, Food and Nature Initiative (2012) *Landscapes for People, Food and Nature: The Vision, the Evidence, and Next Steps.* EcoAgriculture Partners on behalf of Landscapes for People, Food and Nature Initiative, Washington, DC. Available at: http://www.conservation.org/Documents/LPFN-ReportLandscapes-for-People-Food-and-Nature_Eco-agriculture_2012.pdf (accessed February 2013).

Landscapes for People, Food and Nature Initiative Blog (2012) Available at: http://blog.ecoagriculture.org/ (accessed February 2013).

Lebel, J.-M. (2003) *Health: An Ecosystem Approach.* In-Focus Collection, International Development Research Centre, Ottawa. Available at: http://books.google.ca/books?id=FI-H0QFqivMC&printsec=frontcover&source=gbs_ge_summary_r&cad=0#v=onepage&q&f=false (accessed February 2013).

McCartney, M. and Smakhtin, V. (2010) *Water Storage in an Era of Climate Change: Addressing the Challenge of Increasing Rainfall Variability.* IWMI Blue Paper, International Water Management Institute, Colombo.

Millennium Ecosystem Assessment (2005) *Ecosystems and Human Well-being: Synthesis. A Report of the Millennium Ecosystem Assessment.* World Resources Institute and Island Press, Washington, DC. Available at: www.maweb.org/documents/document.356.aspx.pdf (accessed February 2013).

MLTM, PCGG, K-water and WWC (2012) *Water and Green Growth – Executive Summary,* 1st edn, Thematic Publication. Ministry of Land, Transport, and Maritime Affairs, Gwancheon-City and Presidential Committee on Green Growth, Seoul, Government of the Republic of Korea, with Korea Water Resources Corporation, Seoul, Republic of Korea and World Water Council, Marseille, France. Available at: http://www.waterandgreengrowth.org/ (accessed February 2013).

Molden, D., Tharme, R., Abdullaev, I. and Puskur, R. (2007) Managing irrigation systems. In: Scherr, S.J. and McNeely, J.A. (eds) *Farming with Nature. The Science and Practice of Ecoagriculture.* Island Press, Washington, DC, pp. 231–249.

Mulligan, M., Rubiano, J., Hyman, G., White, D., Garcia, J., Saravia, M., Gabriel Leon, J., Selvaraj, J.J., Guttierez, T. and Saenz-Cruz, L.L. (2010) The Andes basins: biophysical and developmental diversity in a climate of change. *Water International* 35, 472–492. doi:10.1080/02508060.2010.516330

Mulligan, M., Saenz Cruz, L.L., Pena-Arancibia, J., Pandey, B., Mahé, G. and Fisher, M. (2011) Water availability and use across the Challenge Program on Water and Food (CPWF) basins. *Water International* 36, 17–41.

NERC (2012) Ecosystem Services for Poverty Alleviation (ESPA). Natural Environment Research Council, Swindon, UK. Available at: www.nerc.ac.uk/research/programmes/espa/ (accessed February 2012).

Nguyen-Khoa, S. and Smith, L.E.D. (2008) Fishing in the paddy fields of monsoon developing countries: re-focusing the current discourse on the 'multifunctionality of agriculture'. Keynote paper presented at the INWEPF-ICID Workshop of the RAMSAR COP10 Meeting, Changwon, Korea, 28 October–04 November 2008.

OECD (2001) Multifunctionality: towards an analytical framework. Organisation for Economic Co-operation and Development, Paris.

Swallow, B. and Meinzen-Dick, R. (2009) Payment for environmental services: interactions with property rights and collective action. In: Beckmann, V. and Padmanabhan, M. (eds) *Institutions and Sustainability: Political Economy of Agriculture and the Environment – Essays in Honour of Konrad Hagedorn.* Springer, Berlin, pp. 243–265.

TEEB (2010) *The Economics of Ecosystems and Biodiversity: Mainstreaming the Economics of Nature: A Synthesis of the Approach, Conclusions and Recommendations of TEEB.* Prepared by Sukhdev, P., Wittmer, H., Schröter-Schlaack, C., Nesshöver, C., Bishop, J., ten Brink, P., Gundimeda, H., Kumar, P. and Simmons, B. United Nations Environment Programme TEEB (The Economics of Ecosystems and Biodiversity) Office, Geneva, Switzerland. Available at: http://www.teebweb.org/wp-content/uploads/

Study%20and%20Reports/Reports/Synthesis%20report/TEEB%20Synthesis%20Report%202010.pdf (accessed February 2013).

UNEP (2011) *Towards a Green Economy: Pathways to Sustainable Development and Poverty Eradication.* United Nations Environment Programme, Nairobi. Available at: http://www.unep.org/greeneconomy/ GreenEconomyReport/tabid/29846/language/en-US/Default.aspx (accessed February 2013).

van der Ploeg, S., de Groot, R.S. and Wang, Y. (2010) *The TEEB Valuation Database: Overview of Structure, Data and Results.* Foundation for Sustainable Development, Wageningen, the Netherlands.

van der Zijpp, A.J., Verreth, J.A.J., Le Quang Tri, van Mensvoort, M.E.F., Bosma, R.H. and Beveridge, M.C.M. (eds) (2007) *Fishponds in Farming Systems.* Wageningen Academic Publishers, Wageningen, the Netherlands.

Vital Signs (2012) Vital Signs. Available at: www.vitalsigns.org (accessed December 2012).

WLE (2012) Agriculture and ecosystems blog. CGIAR Research Program on Water, Land and Ecosystems. Available at http://wle.cgiar.org/blogs/ (accessed December 2012).

World Bank (2009) *Minding the Stock: Bringing Public Policy to Bear on Livestock Sector Development.* Report No. 44110-GLB, World Bank, Agriculture and Rural Development Department, Washington, DC.

Zomer, R.J., Trabucco, A., Coe, R. and Place, F. (2009) *Trees on Farm: Analysis of Global Extent and Geographical Patterns of Agroforestry.* ICRAF Working Paper No. 89, World Agroforestry Centre (ICRAF), Nairobi. Available at: www.worldagroforestrycentre.org/downloads/publications/PDFs/WP16263.PDF (accessed June 2012).

Index

Note: bold page numbers indicate figures, tables and boxes.